Perspectives in Science & Technology

Technical Reports
by
The Science Advisory Council
To The Prime Minister

Vol. II

Under the auspices of
Department of Science & Technology
Government of India

HAR-ANAND PUBLICATIONS
in association with
VIKAS PUBLISHING HOUSE PVT LTD

VIKAS PUBLISHING HOUSE PVT LTD
576 Masjid Road, Jangpura, New Delhi - 110 014

Printed at Sanjeev Offset Printers, Delhi.

Contents

Part 2
Socio-Economic
Areas of Relevance

Foreword

The Science Advisory Council to the Prime Minister was constituted in 1986 as an apex body to advise the Prime Minister on major issues related to science and technology in the country, and also to prepare a perspective science and technology plan for 2001 A.D. The Council was also expected to examine certain matters concerned with the scientific departments, priorities in research and development, technology missions and so on.

The Council has now been functioning for nearly four years. It has studied many issues of direct relevance to science and technology in the country and also several topics with a bearing on the socioeconomic sectors. In carrying out such studies, working groups were set up by the Council and the recommendations and reports of such groups were later discussed by the Council, before finalising the reports. Many of these reports have been disseminated among the scientific community. Although some of these reports may not be comprehensive, they are certainly indicative of the vital issues involved and the action required in the relevant areas. Some of the reports commissioned by the Council related to important topics of futuristic value as well, but it has not, however, been possible to cover all the areas that one would have liked to.

It is indeed gratifying that it has been possible to bring out a special publication containing all the important reports prepared under the auspices of the Science Advisory Council to the Prime Minister. We believe that this publication is likely to be of archival value to the scientific community as well as to policy makers and administrators. We would have very much liked this volume to contain many other vital topics of interest but, unfortunately, it has not been possible to bring out these reports in time for publication.

I would like to acknowledge the support and encouragement of the Prime Minister in the work of the Council. I wish to record the contributions made by the members of the Council whose camaraderie and dedication made the work of the Council a matter of pleasure. I am most thankful to the chairmen and members of the various working groups who produced the reports for the Science Advisory Council. I would like to acknowledge the support of Dr P J Lavakare and the members of the Secretariat of the Council in bringing out the reports. I am also thankful to Dr Srinivas Bhogle of the National Aeronautical Laboratory for his assistance and advice in bringing out this publication.

C N R Rao
Chairman
Science Advisory Council
November 1989.   to the Prime Minister

Editor's Note

This compilation has been divided into two volumes.

The first volume contains (a) the approach paper, prepared by the SAC-PM, on a perspective science and technology plan for the year 2001, and (b) seven other reports on the frontier areas of research and development.

The second volume contains nine more reports broadly classified under the category of 'areas of socio-economic relevance'.

Since the reports have been prepared by different expert teams (please see annexure-2 for the composition of these teams), the style of writing and presentation is not always uniform. Some of these reports have also appeared as individual technical reports of the SAC-PM (the approach paper with the red cover has already been widely circulated), but there are others which are being published for the first time.

Part 2

Areas of
Socio-Economic Relevance

8
Management of Renewable Resources

The S&T inputs, most urgently required for the management of India's renewable resources, relate not so much to one specific issue, but to the introduction of a holistic approach of viewing the system as a whole, and of understanding all relevant interactions, including, of course, the human interactions.

Contents

8
Management of Renewable Resources

Preamble

The Science Advisory Council to the Prime Minister has identified the management of the country's renewable resources as an important area in which the Council would like to suggest some programmes/missions to the Government for its consideration. This is an area of great significance since our ability to sustain the process of national development depends vitally on the health of our resource base of soil, water, vegetation, livestock and genetic diversity. It offers a unique challenge for the problems of management of renewable resources are highly location-specific and have to be largely solved by us through our own efforts. This affords a great opportunity to nurture a truly indigenous scientific and technological effort without being continually worried about whether we are lagging far behind the others and merely reinventing the wheel. Since, by its very nature, the effort will have to ultimately involve people at the grassroots, it is also an opportunity to promote a broad culture of science in our country.

The science and technology effort that underlay the green revolution was just such an endeavour that had to be adapted to our own local conditions and was successfully carried through with many notable organisational innovations. It has rendered us self-sufficient in food and has enabled us to pursue modernisation with a measure of confidence. We are, however, now beginning to be aware of the limitations of this approach, for it was a single minded effort to intensify resource use without adequate consideration for its wider environmental repercussions. The sustainability of growth in our agricultural production is therefore in some doubt, as is our ability to diffuse green revolution throughout the country. Our continued progress therefore depends on a more holistic approach that would have its focus on the health of the resource base. An effort in this direction would require innovations as radical as those that ushered in the green revolution. The purpose of this report is to suggest a programme that would initiate such an effort.　　　　　■

1　　The Problem

A whole variety of problems plagues our renewable resource base:

● Reduction in the percolation of water in the ground, increase in the runoff, deforestation, compaction of the ground due to overgrazing by livestock, inadequate levels of mulch on cultivated fields, excessive growth of built up areas have all led to a lowering of the levels of percolation of rain water in the ground, though little concrete data are available.

● A consequence of increased runoff, along with siltation of river beds, has been the floods. Flood-prone areas in the country have dramatically increased from 19 million ha. in the 1960s to 59 million ha. in 1984. These include areas in entirely new states like Andhra Pradesh, Madhya Pradesh, Maharashtra, Gujarat and Tamil Nadu which had no flood problems before the 1970s.

● Irrigation water is used in a very inefficient manner through flood irrigation. Sprinkler, drip and other systems that could tremendously enhance the efficiency are, as yet, little experimented with and little used.

● Water logging and salination plague vast tracts of land, for instance in the black cotton soil belt in Karnataka, Maharashtra and Madhya Pradesh, and the so-called user lands in Uttar Pradesh. Pushing irrigation projects without adequate provision for drainage is the single most important cause of this disaster.

● Inadequate recharge of ground water and its overdrawal have led to serious problems of lowering of ground water table, with many open wells going dry, for instance, in Gujarat, the Marathwada region of Maharashtra, the Anantpur district of Andhra Pradesh and the Coimbatore district of Tamil Nadu. However, bore wells continue to be dug without any investigation of whether the drawal of water is being compensated for by recharge, and without any measures for enhancing the levels of ground water recharge.

● There has been extensive pollution of surface as well as ground waters. Again no thought has been given to the costs involved; for instance, in the loss of fisheries and whether the acceptance of these costs is justified by the benefits conferred by the activity generating the pollution.

● Deforestation, overgrazing, improper road construction: a number of causes lead to heavy soil erosion losses, estimated at 16 tonnes per ha. of India's land surface. This results in steady depletion of the productive potential of our land that never forms part of any economic calculus.

● Something like 3000 million tonnes of soil is annually deposited in our stream and river beds, and 500 million tonnes in the dams. This cuts down on the irrigation potential and increases flood damage very substantially.

● Our present day agricultural technology would easily allow us to produce enough food from good, fertile lands alone. Nevertheless, large tracts of hill slopes or extremely dry lands, that would be much better left under tree and grass cover, are brought under the plough by subsistence farmers just to eke out a living. The grain production from such lands is often a meagre 3-4 quintals per ha. and the farmers cannot afford any soil and water conservation measures. The result is an enormous burden of soil erosion. But there is no reliable database on the relation between land capability and land utilisation, and no well-thought out programmes for bringing about more appropriate land use.

● It is suspected that with the depletion of sources of organic manure, and the reliance on chemical fertilisers that do not supply micronutrients such as zinc and boron, our agricultural soils are being progressively depleted of their nutrient capital. Again, there is little reliable information on this vital problem.

● Fertile soils have suffered degradation from mine tailings, from brick making, from industrial effluents and through improper use of fertilisers and pesticides. Again there is little systematic information on this problem.

● There has been a tremendous loss of traditional crops, such as leafy green vegetables or tubers and of cultivars of cereal crops, fruit trees and so on, as well as strains of domesticated animals. While a serious effort has been mounted for *ex situ* maintenance of the cultivars of major crops, there has been no attention paid to other plants, e.g., the innumerable mango varieties of the Western Ghats, or to *in situ* conservation.

● With attention focussed on production of cereals and cash crops on agricultural land, and timber and industrial softwood in forest plantations, our needs of fuelwood, fodder, organic manure and small timber have been grievously ignored. The result has not only been the overharvest of plant biomass with degradation of forests and pastures, but also a further chain of consequences such as soil erosion, siltation of reservoirs, reduction in recharge of ground water, drop in ground water levels and so on.

● India supports half of the world's buffaloes and one seventh of world's cattle and goats on one fortieth of the world's land. As the livestock population has increased, grazing lands have shrunk by being brought under the plough and have become more and more overgrazed. Still practically nothing has been done to enhance our fodder resources and change the animal husbandry practices. There is little scientific information on the livestock, under field conditions and constraints, under which they are managed. The grazing pressure has a whole series of implications for destruction of vegetation, compaction of soils, reduction of ground water recharge etc. which are utterly ignored.

● Plant production on the non-agricultural and forest lands of India is totally ignored; so that the productivity levels are far below the potential, often less than 10%. There is little scientific information available, except bureaucratic data on the extent of land under control of Revenue and Forest Departments. Even then, there is little reliable information on the extent of encroachments on Government land. The whole issue of management of these lands needs serious attention from natural as well as social scientists.

A variety of causes underlie these problems. Amongst the most significant of these are:

● Narrow sectoral approaches by the experts and development administrators. Thus irrigation projects have consistently neglected the vegetation cover of catchment areas, and veterinarians the provision of fodder. More complex interactions such as the role of overgrazing by cattle in depleting vegetation in catchments of rivers have been totally ignored.

● Neglect of many vital issues, such as fuelwood and small timber needs or destruction of fisheries by water pollution.

● Neglect of long term considerations, a., for instance, what happens to the fertility of soils under continued application of chemical fertilisers?

● Neglect of traditional knowledge and practices, for instance, on the use of neem as a pesticide or the cultivation of highly nutritious green leafy vegetables.

● Neglect of social implications of resource use; thus, in Karnataka, thousands of hectares of former C and D class revenue lands are proposed to be utilised to grow softwood for industry. Since villagers critically depend on these lands for their biomass needs, the project has met with widespread opposition.

We therefore suggest that the science and technology inputs now most urgently called for relate not so much to one specific issue, such as provision of good drainage under irrigation or genetic upgrading of livestock, but to the introduction of a holistic approach, of viewing the system as a whole, of understanding all relevant interactions, including, of course, the human factors.

2   Role of Science and Technology

Such an approach requires a spatial focus. The selection of a unit in this context is bound to be somewhat arbitrary, for the dynamics of resources in no locality, however chosen, would be free of outside influences. Thus paper mills in Karnataka draw on bamboo from Arunachal Pradesh, and their effluents affect fisheries in Tamil Nadu. From a physical point of view, the topography, as reflected in a watershed/catchment, provides an appropriate spatial unit, as does a village from a social or a mandal panchayat from an administrative viewpoint. An area of a few hundred to at most a few thousand hectares, with a human population of a few hundred to a few thousand people, would provide the right scale for a sufficiently detailed planning and monitoring exercise. We therefore suggest that the choice should be of one or more villages, falling within the jurisdiction of a single mandal panchayat, and with land area corresponding, as far as possible, to one catchment of appropriate order.

Within such a microcatchment the tasks to be undertaken will be:

● Surveying and inventorying the status of resources including land, water, cultivated crops, orchards, livestock, natural vegetation and animal life, as well as human population.

● Investigation of the dynamics of the resource base, for instance removal and replenishment of nutrients from agricultural lands, annual increment and harvest of fuelwood produced on private, village grazing and forest land, and how its harvest is regulated.

● Documentation of folk knowledge and traditions of resource use, for instance, maintenance of village irrigation tanks, systems of intercropping, regulation of grazing.

● Assessment of problems of resource degradation in the light of socio-economic structures and processes.

● Identification of the causes leading to resource degradation; both the triggering factors and the factors sustaining degradation processes.

● Identification and evaluation of corrective as well as preventive technological packages, which may already be available, and a clear definition of those which need to be evolved for a given set of conditions.

● Identification of ways and means of implementing these technological packages within the given social, economic, administrative and legal constraints.

● Communication of the identified package of practices to the local population and the administrative machinery.

● Monitoring of the on-going patterns of resource use, including, especially, the implementation of the recommended package of practices.

Ideally, the execution of these tasks should be a co-operative endeavour of the scientific/technical experts from different disciplines, teachers and students in local educational institutions, local people and their organisations, such as yuvak and mahila mandals and mandal panchayats, and development planners and administrators.

3 Relevant Experience

We can draw upon the experiences of several programmes in the past to work out a framework within which to carry out the proposed experiments. These include, amongst the Government sponsored efforts, the Karimnagar district experiment of CSIR; the watershed development programmes, under the dryland development project of the Government of Karnataka; the comprehensive watershed development programme, of the Government of Maharashtra; the Uttara Kannada microcatchment ecodevelopment programme of the Karnataka State Council for Science and Technology; the Sukhomajri experiment of the Central Water and Soil Conservation Research and Training Institute and the voluntary efforts of the Dasholi Gram Swarajya Mandal.

CSIR had attempted to prepare a comprehensive plan for the development of Karimnagar district of Andhra Pradesh. The basis of this exercise was the use of aerial photographs along with minimal ground checks, leading to the production of detailed 1:25000 scale maps on the following themes:

- Present land use
- Geology
- Hydromorphology
- Forestry
- Soils
- Land resources
- Optimal land utilisation

These are obviously part of the essential data for good resource use planning. However, this is not enough, for the maps provide only static information on resources; their dynamics is equally, if not more, relevant and has to be taken into account. In fact the preparation of optimal land utilisation maps without this base can only be preliminary. Nevertheless, this was a first valuable attempt at looking at, and planning for, good resource use in a holistic fashion. It failed to progress because CSIR ended up acting alone and could not secure the essential cooperation of other organisations such as the Indian Council of Agricultural Research (ICAR).

Sukhomajri, a village in the upper catchment of Sukhna lake supplying water to Chandigarh city, has been the site of an experiment in soil and water conservation by the Central Soil and Water Conservation Training and Research Institute of ICAR. A unique feature of this experiment has been the establishment of a local "Water Users

Association", and the working together of experts and local people.

Dasholi Gram Swarajya Mandal is a voluntary organisation, in the Alakananda Valley of Garhwal Himalayas, devoted to the tasks of eco-preservation and eco-restoration. They have been pioneers in involving local villagers including women and harijans in planning for good resource use of their own locality. Unfortunately there has been little tie-up with either technical experts or Government machinery.

The dryland watershed development programme of the Karnataka Government is an ongoing attempt at looking at soil and water conservation of a watershed as a whole on an integrated basis, involving coordination of activities of Agriculture and Forest Departments. The University of Agricultural Sciences, Bangalore is also actively involved in this programme with an operational research programme in one of the watersheds. Each watershed is being looked after by a multi-disciplinary team, which however lacks some critical inputs such as from the Animal Husbandry Department. There is also little active participation and learning from the experiences of the local population, although there is an active campaign to educate them. This most valuable beginning is now being taken further by a Comprehensive Land Use Management Programme (CLUMP) that pays due attention to the animal husbandry sector and aims to involve the local people more actively in programme formulation and implementation.

The Western Ghats Ecodevelopment Action Research Programme of the Department of Environment is attempting to focus on several identified micro-catchments in the hilly tract of the Western Ghats, and promote programmes of resource planning and development, by simultaneously involving scientific institutions, universities, colleges, schools, local people as well as Government departments. The experiment is progressing well in two micro-catchments in the Uttara Kannada district of Karnataka. Here, very detailed ground surveys, based on mapping by the Survey of India at 1:5000 scale, have been followed by the preparation of detailed action plans by a multidisciplinary team of scientists put together by the Karnataka State Council for Science and Technology. Local high schools and junior colleges have also been actively involved in the collection of data and the preparation of these plans. These plans are now being implemented in consultation with Government development agencies and with the help of ecodevelopment task forces made up of local people.

4   Action Plan

4.1   Short term

This is evidently a mammoth task which calls for a co-ordinated effort of agencies as diverse as agricultural universities, local high schools, mandal panchayats and forest and animal husbandry departments. While it is essential that we must address this task and evolve ways and means of bringing all the actors together, we must begin the effort on a manageable scale and gain experience before venturing on a more massive scale. We therefore suggest that the first phase should concentrate on a small number of catchments of a few hundred to a thousand hectares, selected to represent the various ecoclimatic zones of each state.

We suggest that the Department of Science and Technology of the Government of India serve as a nodal agency to handle the whole programme at the Central level, with the concerned State Councils of Science and Technology being made responsible for the actual execution of the programme in the experimental catchments. The actual choice of catchments would depend on (a) the level of competence of the various State Councils to take up such a programme, (b) the willingness of local Universities and research institutions to participate and (c) the availability of active voluntary agencies like Dasholi Gram Swarajya Mandal to co-ordinate the effort at the local level. These programmes should be linked to the on-going Watershed Development Projects involving ICAR.

The Karnataka State Council for Science and Technology has developed some valuable experience in the execution of such a scientific programme over the last two years. Based on this experience they have formulated a five year programme which provides useful pointers as to how the different State Councils may evolve their own programmes. Appendix-1 provides further details of this proposal from the Karnataka State Council.

An important component of the programme would be the involvement of local educational institutions. In particular the local colleges should be encouraged to include project work on resource inventories, resource dynamics as well as the social dynamics of resource use as part of curricular requirements. Simultaneously, we should promote the development of new syllabi for courses that would emphasise material on resource dynamics and the social dynamics of resource use as the theme around which different subjects are to be taught. This will have to be backed up with a centrally coordinated effort at generating textbooks, workbooks, manuals and other materials for such an approach at college education. The programme proposed by the Karnataka State Council for Science and Technology also visualises involvement in the preparation of such educational material.

The programme should have 5-year guaranteed continuity with full autonomy in using the funds flexibly, without, of course any expenditure on construction, vehicles, other than bicycles or mopeds, expensive equipment or excessive air travel. With such constraints, an amount of Rs 50 crores may be made available to DST.

4.2   Long term

In the long run, of course, we envision such careful detailed local-level planning of resource use to become a norm all over the country, with the exercise being gradually extended to other districts, and with more and more watersheds in each district. Simultaneously, school and college curricula and teaching practices should be revolutionised using resource dynamics and resource use as a central theme, so that what is taught will be related to actual application of science and technology in the real world around the students. Relating modern knowledge and practices of resource dynamics and resource use to traditional knowledge would serve to both enrich our own science and technology and make it more readily accessible to the bulk of the population, thereby helping create a genuine culture of science and appreciation of technology in the society at large.

5   Concluding Remarks

Such an endeavour will necessarily be a peculiarly Indian endeavour. No other country will have an interest in developing such a knowledge base which is relevant to us, and us alone. This strongly contrasts with the science and technology base underlying defence or urban consumer interests. In these areas other countries have strong interests, obliging us to often fall behind and be forced to borrow from others destroying our faith in ourselves. Undoubtedly this is no justification for abandoning serious effort in these areas, rather it calls for focussing them on a limited number of defined programmes so that we can achieve something worthwhile with concentrated effort. Nevertheless, that would still mean a rather limited approach for much of our science and technology effort with no hope of it touching the masses, of generating a broad culture of science that is necessary for us to become a modern nation. On the other hand, the focus suggested here would generate a massive, fully indigenous science and technology effort that would have real relevance not only to solving our pressing problems of bringing about good resource use and creating large scale employment, but also to help develop a broad-based national culture responsive to science and technology.

6   Recommendations

Our ability to sustain the pace of national development critically depends on the health of our base of renewable resources of soil, water, vegetation, livestock and genetic diversity. Unfortunately, our development efforts have tended to stress intensification of resource use without due regard for long term sustainability of the effort and for concomitant degradation of other resources. The consequences have been disastrous; large scale soil erosion, siltation, floods and droughts, waterlogging and salination, lowering of ground water table, gradual depletion of fertility of fields, extensive deforestation and overgrazing by an ever exploding population of cattle and goats. This difficult situation offers a unique challenge to our science and technology establishment because the relevant problems are highly location-specific and have to be largely solved through our own efforts. Since the efforts will necessarily have to involve people at the grassroots as well, it is also an opportunity to promote a broad culture of science in our country.

We believe that the science and technology inputs called for relate not so much to specific sectors as to the adoption of a holistic approach, of viewing the system as a whole, of understanding all relevant interactions, including, of course, the human factors. This will have to be a co-operative endeavour involving not only scientific and technical experts, but also the development planners, local educational institutions and other organisations such as yuvak and mahila mandals and mandal panchayats. The appropriate spatial scale for planning, implementing and monitoring of good resource use in such an effort will be a micro-catchment of a few hundred or thousand hectares corresponding to one or more villages.

We propose that the Government of India launch an experimental five year programme of application of science and technology to microlevel planning of utilisation of renewable resources for a series of micro-catchments representing the different ecological zones of the country. This programme, with a total budget of Rs 50 crores, could be co-ordinated by the Department of Science and Technology of the Government of India and implemented through the State Councils of Science & Technology. Our ultimate aim, of course, would be to extend such an approach to cover the entire country in three phases over the next fifteen years.

Appendix-1

Proposal for an Action Research Project on Environmentally Sound Development Strategies Karnataka State Council for Science and Technology

Contents

1 Executive Summary

The Karnataka State Council for Science and Technology (KSCST) proposes to undertake a five year project on evolving environmentally sound development strategies in five microcatchments representing the different ecological zones of the state. The project would be executed by a multi-disciplinary working group including technical experts, government officials and representatives of the local people. This working group would be supported at the field level by village ecodevelopment task forces. This organisation has been evolved and is working satisfactorily in two microcatchments, in which KSCST has an ongoing programme. The project would focus on a holistic view of the resource base to be developed in a sustainable fashion and in a way such that benefits would accrue to the weaker segments of the society. To ensure replicability the outlay on development schemes would be at the current state-wide level of Rs 250 per head per year. The total outlay for the project, including developmental expenditure which would be incurred in any event, comes to Rs 4.5 crores for a five year period.

2 Objectives

To evolve a methodology for utilisation and development of natural resources that would :

- Result in sustenance of the resource base on a long term basis
- Be based on a holistic view of the resource base
- Take proper account of features specific to the local environment
- Be responsive to the needs and aspirations of the local human communities.

3 Background

Our conventional approaches to development have emphasised sectorally oriented programmes of intensification of resource use, without due regard to long term consequences or side effects. The result has often been a liquidation of the capital stock of resources, and a degradation of the environment. The consequences of such exhaustion and degradation have largely been passed on to the weaker segments of our population, while benefits of development have tended to be accrued to those already better off. On the other hand, our attempts at environmental conservation, such as national parks, have remained isolated from the developmental efforts, and have again tended to exclude people rather than involve them. Appreciating these problems, there is now a move to reorient the planning methodology towards patterns of economic development that would use resources in a sustainable fashion, be in harmony with the ecological endowment and pass on the benefits of development to the weaker sections of the society. This coupled to the concern to ensure that development programmes indeed do get implemented as planned, leads to the concept of "environmentally sound development strategies". Given the tremendous diversity of our environmental regimes and social organisations, it is evident that such environmentally sound development strategies would have to be location-specific. A highly decentralised effort, sensitive to the local situation, would therefore be necessary to evolve and implement such strategies. The current focus on district level planning is part of the attempt to initiate such an effort. The present project aims to develop the methodology for evolving such decentralised, people-oriented and environmentally sound development strategies at the level of the most basic planning unit, namely a microcatchment.

4   Methodology

4.1   The Spatial Framework

The theme of environmentally sound development strategies is equally appropriate for a whole range of ecosystems, be they predominantly forested tracts inhabited by tribals, tracts under rainfed or intensive irrigated cultivation, tracts under extensive mining, or industrial townships. Most of our country, and especially the vast majority of its less developed regions, is however under rainfed agriculture or forests and waste lands. The present project is therefore confined to representative examples of these kinds of ecosystems of the state of Karnataka. The examples have been so chosen as to cover the different ecological zones of the state, namely, the high rainfall coastal tracts, high rainfall hill tracts dominated by horticulture, forested tribal hill tracts, semi-arid plateau region supporting large scale sheep rearing, and northern black cotton soil tracts under extensive rain-and tank-fed cultivation.

Ecologically, the most appropriate planning unit for a forested or agricultural ecosystem, especially in the undulating terrain characteristic of most of the state of Karnataka, is the microcatchment (often used synonymously with a watershed), i.e., the region over which water flows into a stream of second or third order. Such a unit need not correspond to an administrative unit such as a revenue village or a mandal panchayat. This would pose some problems that would have to be resolved. We have however decided to stick to the natural unit of a microcatchment for the exercise proposed here.

4.2   The Organisational Structure

Over the last fourteen years, the KSCST has successfully developed the system of executing its projects through multidisciplinary working groups comprising scientists/technologists, Government officials and representatives of other sectors such as industry. This working group is convened by a technical expert who serves as the Principal Investigator and supervises the work of assistants employed specifically for the project. The KSCST staff help in the co-ordination of the project and the liaison with the various state and central government and other agencies.

With the Zilla Parishads assuming full responsibility for district level development planning in Karnataka, KSCST is working out ways and means of providing science and technology inputs to this effort. Such inputs may be provided through the establishment of science and technology advisory groups linked to the Zilla Parishads' Standing Committees on planning. These groups would prove very useful for the execution of the project being proposed here.

The organisation of the present project involves a much greater range of interactions with different elements of the society. (see *Fig. 1*). Considerable experience in organising such interactions has already been obtained through a KSCST project on integrated development of selected microcatchments in the Bedthi-Aghanashini river valleys of Uttara Kannada district initiated in 1986. This project is being co-ordinated by a local voluntary agency, Sahyadri Parisara Vardhini, with technical backing from the Indian Institute of Science, Bangalore and the University of Agricultural Sciences, Dharwad. Sahyadri Parisara Vardhini has promoted the establishment of village level ecodevelopment task forces in the catchments. There are now very regular, generally monthly, interactions of the technical experts, village ecodevelopment task forces and the project staff. Sahyadri Parisara Vardhini and the project staff are also helping in liasing with the Zilla Parishad, the Government officials and other institutions such as banks and co-operative societies. The project proposed here is being organised on the basis of this experience; documentation of this experience would also be a part of the proposed project.

4.3 The Planning Process

The process of planning for environmentally sound development of the identified localities would include the following steps (see *Fig. 2*).

● *Identification of the parameters relevant to the project*   These would range over a variety of parameters, such as rainfall, ground water table, standing biomass and productivity of natural vegetation, land use and cropping pattern, livestock composition and grazing systems, requirements of cooking energy and thatch, the level of education in the local population, functioning of local institutions such as mahila mandals, mandal panchayats or co-operative societies, conflicts over the use of common property resources, availability and implementation of various development schemes of the Government etc. These would be identified by the

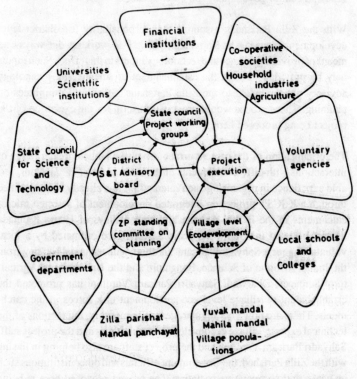

Fig. 1 Organisational structure of the project on evolving environmentally sound development strategies.

Project Working Group set up by the State Council in consultation with local people and institutions, government officials and other technical experts.

● *Data collection* This would include physical surveys, collection of household data, collation of secondary information available with the Government and other sources etc. This task would be carried out by the project technical staff in active collaboration with students and teachers in the local educational institutions and local voluntary agencies.

● *Data processing* This would be the primary responsibility of the project technical staff which would continue to interact with local educational institutions and voluntary agencies.

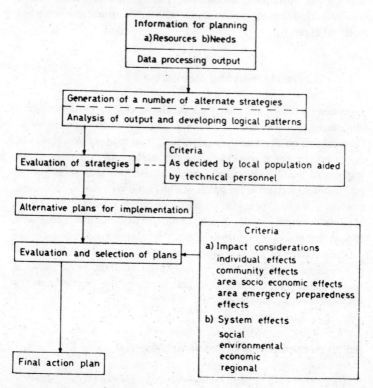

Fig. 2 The selection process.

● *Information* All of this would yield information vital to planning in forms such as resource flow diagrams. These would pinpoint the current imbalances in the system and causes that trigger and sustain such imbalances. It would also suggest various options as to how such imbalances may be corrected and environmentally sound development strategies elaborated.

● *Dialogue with local population and Government officials* The various options, along with relevant selection criteria, would then be put before the local population, largely through the medium of village ecodevelopment task forces that would have members drawn from all strata of the village population, with special emphasis on the weaker segments most heavily dependent on common property resources. This feedback may help to generate further options which would then be elaborated, if necessary with the help of further investigations. The ultimate choice of the development strategy with proper prioritisation should ideally rest with the local village population, and be made in consultation with the Government officials and technical experts.

4.4   Implementation and Monitoring

To ensure eventual replication, the level of financing for the development programme should be at the level of expenditure normally incurred by the Government, which presently stands at Rs 250 per head per year. However, in this programme it would be desirable to place the entire sum at the disposal of the Working Group which would then arrange to get the programmes executed in consultation with the Government agencies, and employing the local people to the maximum extent possible. The process of implementation would include an element of continual monitoring, with enough flexibility to change the specific programmes if so warranted by the experience (see *Fig. 3*). As with planning, monitoring would be the responsibility of the Working Group, and would be carried out by the technical staff employed for the project in collaboration with local educational institutions and voluntary agencies.

4.5   Generation of Instructional Material

KSCST has already acquired substantial experience of evolving environmentally sound development strategies based on its experience in the

Uttara Kannada district since 1986. As the project continues, further experience will be gathered. All of this would be very useful to other groups wishing to undertake similar activities, and the State Council would generate manuals, workbooks, mobile exhibitions, slide shows and other instructional materials to document its entire experience.

4.6   Benefits Expected

The project would directly benefit the weaker segments of the society heavily dependent on common property resources in the chosen localities. These would include the Soligas, a scheduled tribe of BRT hills in Mysore district, and scheduled caste populations especially involved in basketweaving in other project areas. The project would also help develop the methodology highly relevant to decentralised district level planning.

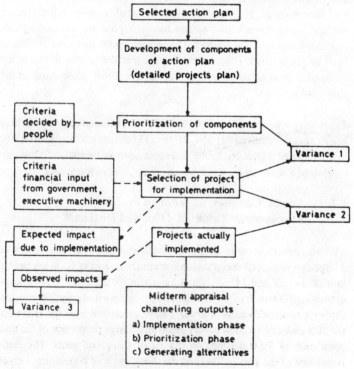

Fig. 3 The evaluation process.

5   Profiles of the Identified Project Components

5.1   Gillesugur Complex of Villages in the Raichur Taluk of Raichur District

This watershed represents the black cotton soil tract of Karnataka with an annual rainfall of 600 mm. The area lies between latitude 15° 55' and 16° 15' N and longitude 77° 20' and 77° 30' E at an elevation of 390 m. The area proposed to be covered extends over 10870 hectares with a population of 16000 and encompasses 14 villages. The crops cultivated include jowar, groundnut, paddy, navane, toor, green gram and bajra in the kharif season and wheat and cotton in the rabi season.

The whole region, earlier, had a well developed system of soil conservation measures including a barrier called Kadagamma and a small pond for every 2 to 3 hectares. There are also 12 larger tanks in the locality most of which have been silted up. The focus of this project would be on reviving a proper soil and water conservation system for the region and developing tree vegetation and fodder resources. The project would also have to address itself to the conflicts inherent in the fact that the people living in the catchment areas of tanks are not directly benefited, while those in the command areas are.

The project would be spearheaded by INGRID, a voluntary agency working in Raichur taluka since 1981. INGRID has 15 full time staff members active in the fields of education, health and community organisation. It has also conducted a study on the status of irrigation tanks in Karnataka.

5.2   Airani Complex of Villages in Ranebennur Taluk of Dharwad District.

This watershed represents a hilly dry tract dominated by red soils and with a large component of sheep keeping. It covers an area of 6000 hectares at latitude 14° 23' and 14° 50' N and longitude 75° 20' and 75° 50' E at an elevation of 600 m. It receives an annual precipitation of only 500 mm. and supports some cultivation of jowar, cotton, groundnut and ragi. However, the area under cultivation is very small and a large proportion of the total population of 7500 depends on rearing of sheep and goats. The cattle population of the area is 3000 and the sheep and goat population is about 10,000. Apart from animal husbandry, the people also depend heavily on spinning, basket making, leather working, fishing etc. There is a substantial amount of migration of people especially of the nomadic shepherds.

The project would focus on good soil and water management, afforestation and fodder development on the hilly lands, better practices of dryland agriculture and development of agriculture and animal husbandry based processing industries.

This project will be spearheaded by India Development Service (IDS), a voluntary organisation working in Ranebennur and Dharwad taluks of Dharwad district since 1979. IDS has a full time staff of 60 working on programmes of health, animal care, social forestry and rural handicrafts.

5.3   Soliga Tribal Settlements of Biligiri Rangana Temple Hills in Mysore District

The selected watershed of 6000 hectares lies between latitude 11° 50' and 12° 10' N and longitude 77°0' 77° 30' E at elevations ranging over 680 - 1200 m. The area receives about 1200 mm of rainfall a year and has a good cover of moist deciduous forest. Apart from 6 tribal hamlets, housing 2600 Soligas, the remaining area is a reserved forest. There are 500 cattle in the area and the tribals practice small scale agriculture with ragi and maize as the major crops. The region supports a good population of wild animals including elephants.

The tribal economy has been seriously dislocated following the banning of their traditional practice of shifting cultivation and hunting of wild animals. The productivity of their settled cultivation is low and is further depressed by damage by wild pigs and elephants. The tribals are heavily dependent on minor forest produce collection and labour in the employment of Forest Department.

A whole new system of resource use needs to be organised with better cultivation practices and the development of live fences to keep out wild animals. There are also good possibilities of cultivation of medicinal plants, bee keeping, fish culture etc. The forest area could be enriched by planting species which produce valuable minor forest produce. The tribals could be organised to harvest and market these produce in an effective fashion.

This project would be spearheaded by the Vivekananda Girijana Kalyana

Kendra, a voluntary organisation active amongst the Soliga tribals. VGKK is primarily engaged in health care and education with a full time staff of 15. It has organised Soliga Abhivridhi Sanghas in each of the 90 tribal hamlets in the Mysore district.

5.4 Masur-Lukkeri Cluster of Villages in Kumta Taluk of Uttara Kannada District

This is a cluster of 9 settlements on an island in the estuary of the river Aghanashini at latitude 14° 28' N and 74° 23' E longitude. at an elevation of 0-75 m. It represents the coastal ecosystem with lateritic soils and a precipitation of 3200 mm a year. The total area of the watershed is 360 hectares supporting a population of 4000 people and 730 cattle. The main crops are fresh water and brackish water paddy and coconut. There is also a considerable amount of fishing and prawn capture in the estuary. The centre of the island is a barren hillock that is now being afforested as a part of the ecodevelopment project initiated in 1986. There were extensive mangrove forests in the brackish water areas; these have been largely destroyed.

There are serious problems of fodder shortage, salination of the brackish water paddy fields and acute shortage of drinking water in the summer. The project is focussing on afforestation and fodder development, water conservation, alternative systems for the management of brackish water paddy lands, including the development of mangrove vegetation along bunds.

The project would be spearheaded by Sahyadri Parisara Vardhini, a voluntary organisation with a full time staff of 8 project personnel. SPV, established in 1984, is principally involved in ecodevelopment and environmental awareness activities.

5.5 Sirsimakki- Mundagesara Cluster of Villages in the Sirsi Taluk of Uttara Kannada District

This watershed covering an area of 400 hectares lies at latitude 14° 35' N and longitude 74° 48' E at an elevation of 600 m. It represents Western Ghat hill tracts with an annual precipitation of 2500 mm. It has a population of 700 people and 450 cattle. Arecanut and paddy are the main crops but the bulk of the land is forest land assigned for the use of orchard owners. There has been substantial overutilisation of these forest lands. There are also serious problems of overgrazing. The focus of the project is on soil and water conservation, afforestation and fodder development. This project is also in progress since 1986 and is coordinated by the Sahyadri Parisara Vardhini.

The project components 4 and 5 in Uttara Kannada district are ongoing projects, and would provide excellent experience in monitoring and help generate instructional material.

6   Financial Requirements

The funds for these projects would be firstly required for actual financing of the developmental activities and secondly for the planning and monitoring activities and for the development of instructional material. The developmental expenditure will be at the rate of Rs 250 per head per year for population in the catchment areas. Since the planning for two of the watersheds is already completed, the expenditure for those catchments would begin from the first year itself and for other areas it would begin from the third year. The developmental expenditure would thus come to about Rs 80 lakhs per year. This development expenditure will in any case be incurred during the course of normal development. The other expenditure including salaries for 30 project assistants at an average of 6 per watershed, working expenses and preparation of educational materials is estimated at Rs 15 lakhs per year. Thus the total project outlay for a five year period comes to Rs 4.5 crores.

7   Phasing of the Project

ACTIVITY	0 1 2 3 4 5 6
1 Generating a database	1.5 Yrs
2 Equipping the target group with skills for selecting options	2 Yrs
3 Implementations	3 Yrs
4 Impact assessment	1.5 Yrs
5 Developing pedagogic material	5 Yrs

9
Health Care

Population growth is a crisis which threatens the entire health care development programme, and even the very future, of our country.

Contents

9
Health Care

Preamble

The guiding principle in respect of health care has been provided in our Constitution in the following words: "the elimination of poverty, ignorance, and ill-health", and the Constitution directs the State to regard "the raising of the level of nutrition and the improvement of public health as among its primary duties, as also securing the health and strength of workers, men and women, especially ensuring that children are given opportunities and facilities to develop in a healthy manner".

Even before Independence, the Bhore Committee (1946) recognised that the large amount of preventable suffering and mortality in India was mainly the combined result of inadequacy of provision in respect of environmental sanitation, nutrition, health education and medical services. It emphasised integrated primary health care, social and preventive aspects of medicine and health care, people's participation and strong inter-sectoral co-operation.

Over the years, through the mechanism of planned development, significant achievements have been made in improving the health status of our people. The life expectancy has increased from 27 years in the 1940s to 56 years today. Small pox and plague have been eradicated, and malaria and cholera are at least partially contained. The infrastructure for health care delivery has been developed throughout the length and breadth of the country. A variety of national health programmes against tuberculosis, leprosy, malaria, filaria, blindness etc. have been instituted. Prophylactic immunisation, prevention of vitamin A and iodine deficiency are also making headway.

In spite of all these developments the overall health scene in the country leaves much to be desired. Thus, in 1983, the National Health Policy Statement commented: "in spite of such impressive progress, the demographic and health picture of the country still constitutes a cause for serious and urgent concern. The high rate of population growth continues to have an adverse effect on the health of our people and the quality of their lives. Thus mortality rates for women and children are still distressingly high..."

The conditions even today, as per the mid-term review of the Seventh Plan, continue to be a cause for concern. There is unabated population growth of around 2 per cent per annum, infant mortality still is unacceptably high at 95 per 1000 live births. There are approximately 10 million active cases of tuberculosis, and one third of the world's leprosy patients are in India. Malaria, filaria, kala-azar, diarrhoeal disease and acute respiratory infections, specially amongst children, are still responsible for serious morbidity and mortality. To this may be added the health burden usually associated with development: cancer, heart disease, strokes, traffic accidents and diseases due to exposure to environmental and occupational pollutants have already acquired epidemic proportions.

Reviewing this scenario, a more critical evaluation of the existing strategies and structures of the health care system in the country, with a view to make specific recommendations for future action, was undertaken. It is hoped that these recommendations will be of use to the planners of health care delivery in the country.

There are enough rays of hope on the horizon which indicate that, given the will and commitment, coupled with better management and affordable inputs, it is possible to achieve our goals in the foreseeable future, as has already been demonstrated in several areas in the country. ◼

1   Introduction

A Committee on health care needs was constituted with the following terms of reference:

- To review the existing health care delivery system in the country, and its strength and weakness.
- To project the health care needs for the turn of the century.
- To suggest ways and means to meet the expected challenges.

The Committee met on several occasions and, besides their discussions during the meetings, also submitted background papers.

This document has been drafted by the Committee in all humility, being very well aware of the many laudable and significant achievements since independence in the area of health. The intention is not, in any way, to belittle what has been, and continues to be done today, but is rather motivated by an impatience to progress even faster as we firmly believe that India can achieve much more.

In India only about 3.7% of our GDP is allocated to the health and family planning sectors taken together. The health sector is an important social sector with direct implications not only for the quality of life, but indirectly for national development and productivity. It is important both to ensure that increased resources are made available to this sector and to see that whatever resources are provided are equitably distributed according to national priorities, and with emphasis on the tenets of primary health care, for the benefit of the people, not forgetting the rural masses and the poor.

The primary health care concept is not restricted to the provision of services and infrastructure at the periphery alone; health care delivery should be seen as a continuum, a spectrum that ranges from basic care delivered by the community itself at one end to the most evolved tertiary care facility at the other extreme. *The entire chain is needed if any part of the system is to work at optimal efficiency.* It is therefore essential that the various segments do not function in isolation, but as a well-knit organic whole. While some attention is being paid to planning for the primary and secondary care facilities, the tertiary care sector seems to be characterised by unplanned and unregulated growth. The apical institutes are no longer functioning as referral units but are rapidly being forced into the role of general hospitals. These institutions are also having to contend with the effects of a rapidly eroding autonomy.

It is important to give the community an active role in the health care scenario; this people's participation must not be merely passive acceptance of health care but should progress towards the ideal of an active partnership and a dynamic decision making role for the community. It is unfortunate that the health system has, by and large, treated the concept of community participation as merely a method of helping the health systemachieve its targets of health care delivery rather than as a method of health empowerment of the people.

Despite the fact that most health parameters have exhibited a curious inertia, notwithstanding the many plans and schemes implemented in the country, there are islands of hope that have shown unequivocally that, with better management, and some affordable inputs, *it is possible to achieve most of the national targets for Health for All by the year 2000 AD,* except probably the control of population growth which would require a different approach. And these examples are not the prerogative of any one sector alone; they range across various domains from Jamkhed (NGO), CMC, Vellore (private) to AIIMS, Ballabgarh (government).

On the other hand the mid-term review carried out by the Planning Commission for the Seventh Plan indicates the inability of the existing plans and policies to meet even the limited targets in most areas of health care delivery. Leave aside the ultimate goals in terms of improvements in the health standards of the country, even fiscal targets for infrastructural and manpower development are unlikely to be achieved. This is compounded by the less than optimal output from existing facilities. In many cases it is not the dearth of funds but that of proper utilisation that is responsible for such failures. The underlying causes include the lack of managerial and epidemiologic inputs into the system.

The importance of areas that impinge on health, the 'non-medical' subjects such as water and environmental sanitation, nutrition, the socio-cultural perceptions of the community, women's status, literacy etc. have perhaps not been adequately realised, both by the medical sector and the administration. Most improvements in the health status of communities have resulted either from general 'non-medical' developments or by public health action. Curative care does not normally influence the health status of the community though the recipients of effective curative care are motivated towards accepting preventive and promotive advice.

While planning it is not only necessary to look at the current needs and deficiencies but also to consider the likely scenario in future. Some of the important changes expected in the demographic and health parameters by the end of the century are:

● Unabated increase in population would require provision for health care on a much larger scale than hitherto envisaged. Population growth is a crisis which threatens the entire development programme and even the very future of our country.

● As a result of increasing population, coupled with an improved child survival since independence, the pool of young people in the reproductive age group will increase manifold. This will further aggravate the population growth situation.

● Changing demographic patterns with respect to age, would require provision for health care for nearly 100 million aged people in the community. The aged as a group have not been taken into consideration in earlier plans.

● Continued migration of rural population will carry the obvious risk of creating more slums in urban areas. Strategies will have to be evolved for preventing this and, at the same time, more attention would have to be paid to urban health than has so far been planned.

● Changing morbidity profiles are likely to result from the increasing average age, urbanisation and industrialisation on the one hand

and hopefully successes in programmes for eradication or control of common communicable disease on the other. This would require planning for prevention and management of non-communicable disease such as heart disease, cancer, traffic and industrial accidents, the effects of environmental degradation and pollution etc..

Health care can be provided and people can reap the benefits of scientifically sound, but low cost technology, as is evident from several examples in our country. It is for these reasons that we deliberated and arrived at the conclusions discussed in the following sections.

2 The Health Care Delivery Infrastructure

The general strategy of having multipurpose workers at the periphery, and the creation of sub-centres, primary health centres, community health centres and district hospitals is strongly endorsed. However, it has become increasingly clear that a functional system exists more on paper than on the ground, even though the nation can boast of having established an extensive network of manpower and physical infrastructure. It is necessary to strengthen, sustain and efficiently utilise this investment (see Table 1).

One link that had been established for primary care was the village health guide. Unfortunately, a vacillation about this position has created a cadre of 3,87,436 dissatisfied people in the rural areas. It is recommended that the salaries of the concerned people should be paid upto a suitable cut-off date, such as the 31 March 1990, and decisions

Table 1

Health infrastructure established

	# in 3/89	# for 1990
Sub-Centres	120767	130000
PHCs	18811	21666
CHCs	1631	2708

then taken about the continuation or otherwise of the scheme. In general, we agree that the concept of having community health workers (CHW) is useful and a beneficial adjunct to the primary health care infrastructure. Whether the CHW takes the form of the village health guide, is a male or female, or follows the information, education, communication (IEC) training plan and uses link persons who are totally unpaid, is a matter of detail, and needs to be carefully evaluated.

The multipurpose worker scheme is sound and forms a crucial link in the chain. However this scheme has not been implemented in toto. The ratio of male workers to female workers is supposed to be 1:1; the actual ratio is between 1:1.5 and 1:2. As the states are supposed to pay a large part of the salary of male workers, while female workers are centrally supported, the male female ratio is distorted in almost every district in the country. A recent calculation for a district in Madhya Pradesh (Table 2) highlighted a 50% shortfall of male workers and a 30% deficiency of female workers. This is a randomly chosen example but is, by no means, an atypical situation in the field.

Table 2

Multipurpose workers, Chindwara district

Workers	Multipurpose Workers		
	Male	Female	Total
GOI norms*	309	309	618
Sanctioned	224	224	448
In position	153	224	377
Shortage from			
a) Sanctioned	74	0	74
	(33.0%)		
b) GOI norms	156	85	241
	(50.1%)	(27.5%)	

*Norms are calculated on the basis of the current estimated population and the recommended sub-centre for 5000/3000.

Another anomaly exists because the male worker is still funded out of a plethora of projects and schemes and thus have different pay scales and different allowances (even though they are supposed to be doing the same work). There does not seem to be any reason to continue to have designations such as 'small pox worker' or 'vaccinator' when it is now over a decade since small pox was eradicated in the world. Another similar anachronism is the continued existence of the 'trachoma worker' on the pay and accounts lists while the functions were given up long ago.

Wherever male workers are available, and are actually functioning, they are treated as uni-purpose workers. The malaria worker is a case in point. The entire national programme is maintained as a de facto vertical programme to the detriment of both the malaria programme and other programmes in the field.

The experience in the states of Bihar, Madhya Pradesh and Uttar Pradesh has shown a remarkable increase in the acceptability of health care and the early trend suggests that the functioning of the national health programmes, such as malaria and family planning, is also improving where the IEC Training Plan has been implemented. This programme of the Ministry of Health and Family Welfare, being implemented in the field by the Centre for Community Medicine (AIIMS), aims at improving the total functioning of the health system using the existing infra-structure. The inputs provided are managerial, coupled with regular training, health education and community involvement through totally unpaid volunteers. If this scheme lives up to its early promise it will be worth emulating widely.

It is strongly recommended that the male workers be brought under one scale of pay and the different designations, that are being kept alive only because of the different pay heads, be abolished. Some effective steps need to be taken to ensure that the MPW (male) is actually appointed and then begins to work in tandem with the female worker. The actual MPW scheme envisaged functional changes upto the district level. Steps should be taken to implement the scheme in its entirety.

The female MPW is paid entirely from central funds, the sub-centre medicines are similarly part of the family welfare planning budget and thus 100% centrally sponsored. Even a substantial portion of the salary of many male workers are paid by the centre (40% of the salary of the largest category of male workers, the malaria workers, is centrally budgeted; 100 % of the salaries of the erstwhile family planning social worker is central etc.). One option suggested is that *the Centre should take over the entire responsibility for the pay of all the sub-centre staff.* The existing pay slots thus freed for the state governments could be used for additional inputs into the rural health services.

We are in full agreement with the concept of the primary health centre (PHC) and feel that this institution, which is planned as the backbone of the health care delivery infrastructure, needs to be strengthened and supported. While the population of 30,000 to be covered is small enough to permit intensive coverage, it is recommended that as far as possible a minimum of two doctors be posted at each PHC so as to not only facilitate attention to both field duties and curative care at the PHC itself, but to also remove professional isolation.

The upkeep and general maintenance of the equipment and physical infrastructure of the health department, especially in the rural areas, is really regrettable. It is important, for reasons of long term economy, staff morale and client confidence, to ensure much greater attention to this aspect. Once the PHC system starts functioning effectively, steps can be taken to enhance the functions and capacity. An example for future consideration is the area of mental health.

The next link in the chain of health infrastructure is the community health centre (CHC). We consider CHCs to be most crucial and yet, today, they are completely distorted. No firm pattern has emerged for the CHC. It was emphasised that this level was not only envisaged as being the first referral point for curative care but was also supposed to similarly provide the direction and leadership for public health and preventive/promotive care for the four PHCs under its jurisdiction. To do this it is essential that, in addition to the four specialists already approved, steps should be taken to include a person trained in community medicine/epidemiology. It is this person who will be responsible for preventive and promotive health services for the entire block and exercise a supervisory function over the peripheral PHCs. He will also be responsible for the health information system and the block level planning.

Other services such as dental health facilities, a better diagnostic service etc. can be extended to the CHCs in a phased manner.

The district health set-up has also been planned adequately. However, even though, on paper, the multipurpose scheme has been extended throughout rural India, in actual fact there have been no changes in the division of functions of the district level officers. Under the scheme it was envisaged that the available senior supervisory staff, such as the deputy chief medical officer and the programme officers, will divide the district between them and each will be responsible for all the functions of their PHCs and CHCs. This has not been done and has resulted in inefficient functioning and the perpetuation of the vertical programmes. Another unfortunate result is a dichotomy of function and control for the multipurpose worker. *It is recommended the the functions of the district level officers be modified so as to divide the PHCs between them, each of them taking total responsibility for their area.*

If Maharashtra is considered to have a good rural health care system, we must learn the lesson from the fact that the three ways that Maharashtra differs from the other states are: (a) a mandatory public health training for district health staff, (b) a good health information system and (c) the involvement of the Zilla Parishads in health care.

The last link in the chain are the medical colleges and the regional and national institutes. These have been dealt with in greater detail in the section on medical care. It must be emphasised, however, that unless an effective referral system is established from the most peripheral point to successively more evolved institutions, the credibility of the primary care system would continue to suffer.

We discussed the concept of having compulsory national service for doctors. It was strongly agreed by all that national service was desirable. Even a two-year period of compulsory national service would make about 30,000 doctors available to the system. However the government could take such a step only if it met two prerequisites: (a) The requirement would only be acceptable if it was mandatory for all, regardless of any other consideration, and (b) it should apply to all professional courses, and not only to medicine.

Even pending the establishment of a system of national service, it is important to have a policy regarding the posting of doctors to peripheral areas. A pre-determined and fixed transfer policy which allows all doctors joining service to have the assurance that every one has to rotate through all categories of health care facilities will remove the fear that some people, for reasons of luck or influence, will never put in service in distant rural areas, while others not so favoured may be stuck for long periods in unpopular postings.

Active steps should be taken to make the health services relevant by ensuring that a block level planning exercise is carried out at regular intervals, preferably annually. The people must be involved in this activity and in the delivery of health services. One method would be to link the health care delivery system with the local Panchayat structure for various activities such as planning, selection of CHCs, setting local priorities for health interventions, organising local meetings etc..

Urban populations today are dependent upon a few general hospitals and tertiary care facilities for curative care as far as the government sponsored health care facilities are concerned. The result is that tertiary care institutions no longer function as referral institutions. A major portion of their manpower time is spent in general care that could be managed just as effectively at lesser structured health facilities. We feel that there is an urgent need to establish an urban primary health care infrastructure. This would not only bring health care closer to the people, especially those members of the community who live in less advantaged circumstances, but would also improve the functioning of the tertiary care facilities.

3   Family Welfare Planning

The mid-term family welfare appraisal of the Seventh Five Year Plan is alarming. It indicates that the rate of our population growth continues unabated, in spite of the enormous amount of work done with social, S&T and modern communication inputs. We therefore face the frightening prospect of an uncontrolled chain reaction into the next century (please see page 30 of this compilation for a more detailed discussion on this subject).

We strongly feel that it is this area that needs the greatest attention and radical change, and that too at the earliest possible time. We are concerned by the fact that, in spite of the great priority placed on this subject, even at the expense of distorting the entire health care set-up, the benefits derived were far less than had been anticipated. *Without definite changes in the policy it is extremely unlikely that the country would be able to arrest the runaway population growth*. The question of India achieving a NRR of 1 by the year 2000 now seems a chimera at the present rate of progress. This is borne out by the latest projections made by an eminent group of scientists for the planning commission (please see *Fig. 1* and Table 3).

If the data of Table 3 are taken together with estimates of the

Table 3

Likely levels of vital rates

| Period | Likely Vital Rates | | | Likely Life Expectancy | |
	CBR	CDR	GR	Males	Females
1981-86	33.6	12.3	21.3	55.6	56.4
1986-91	30.9	10.8	20.1	58.1	59.1
1991-96	27.5	9.4	18.1	60.6	61.7
1996-01	24.9	8.4	16.5	62.8	64.2
2001-06	23.0	7.8	15.2	64.9	66.3

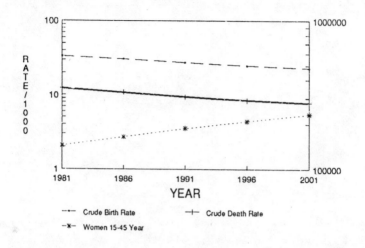

Fig. 1 Likely level of vital rates (1981-2001)

population structure it will be seen that the base of women in the fertile age group will continue to rise in actual numbers in spite of almost any reasonable projection of family planning achievements that we can expect today (please see *Fig. 2* and Table 4).

The separation of family welfare planning from health services may have had a justifiable rationale at one time. Today this division extends to all levels ranging from the Planning Commission (there are

Table 4

Population projections (revised), India, 1981-2006

| Year | Projected population ('00) | | | % | No. females 15-44 | |
	Persons	Males	Females	Urban	Total	Married
1981	6851590	3543843	3307747	23.31	143857	115776
1986	7623597	3934483	3689114	25.51	164373	129161
1991	8429786	4342410	4087376	27.91	187694	144081
1996	9230336	4747619	4482717	30.45	209150	157622
2001	10023848	5148106	4875742	33.16	232757	173148
2006	10814582	5546983	5267599	35.93	255697	187730

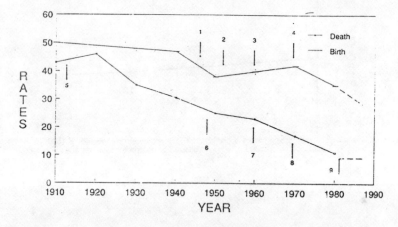

Fig. 2 Birth and death rates in India (1910-1990). [1] partition and World War II; [2] postwar baby boom; [3] increased couples; [4] expanded family planning services; [5] World War I and influenza; [6] independence, and more medical colleges; [7] national programmes; [8] expansion of primary health centres; [9] EPI.

separate steering committees for health and family welfare planning!), the Ministry, to even the peripheral workers who owe allegiance either to one or the other stream. This is to the detriment of both. Because of the divorce from health, family planning is not seen as either a peoples' movement nor as something for the benefit of the community. Health, on the other hand, has had to take a back seat because the targets for the staff relate to family planning performance and the assessment of the staff, ranging from the civil surgeon (chief medical officer of health) to the peripheral worker, is based on 'cases' enrolled for fertility control. Studies have shown that the villagers perceive the PHC and the sub-centre as family planning institutions and not for promotion of health, not even curative care.

Health and family welfare should be amalgamated into one comprehensive, integrated and inter-dependent activity (and department) as it was in the time of Pandit Jawaharlal Nehru.

The system of monetary incentives has mutated from what was originally a compensation for loss of wages, into a system of virtually buying 'cases'. This, coupled with the practice of personal targets, results in all types of unhealthy practices where different persons and different departments are vying with each other and virtually squabbling over patients who are seen as 'cases' rather than as persons.

Monetary awards to States has also resulted in an emphasis on numbers instead of paying attention either to the quality of care or even to the appropriateness of those subjected to surgery. Another unfortunate outcome has been that the pressure caused by competing for awards has resulted in statistics that have little utility for planning or forecasting population trends.

The system of monetary incentives for contraception needs to be abolished. At the personal level it is fostering the impression that by accepting contraception, the individual is obliging the health system; at the provider's level, this approach, coupled with the invidious personal target practice, is serving to dehumanise the delivery of services; at the State level it is leading to concentration on reporting quantitative tabulations with little or no concern for quality or even, in some cases, veracity!

There still seems to be a lack of internal consistency in the policies of the government. For example if a person in government service with two children opts for family planning he/she gets an incentive in the form of one increment in salary; however if a person takes the same decision after just one child, the incentive is not available. Certain segments get a children education allowance. This is given *pro rata* for each child and does not stop after the second child, a clear financial incentive towards having a large family. Even development aid to villages is population linked: roads for a population of 1000, primary schools for 3000, hand pumps for 250 etc., any thinking sarpanch would naturally want as large a population as possible!

It is recommended that such anomalies be removed and *village development inputs be linked to family planning performance and not to the strength of the population.*

Laparoscopic ligations have also proven to have a totally unacceptable number of unfavourable outcomes. Not only is the failure rate very high but avoidable morbidity and even mortality has resulted from its use under less than optimal conditions. This has had a very adverse impact on the acceptability of the whole programme.

Laparoscopic sterilisation should not be allowed at levels below that of a community health centre, and that too to be done by a specialist. At the PHC and similar levels the method of choice should be the mini-lap. It would also be useful for government to get a detailed inquiry conducted into the issue of laparoscopic ligations.

Studies have shown that people from both rural and urban areas are generally aware of the family planning programme, and even of the the more widely used forms of contraception. Research has also highlighted that a very large number of couples have tried and then given up some method of family planning. This points to two things. Firstly we need to continue the search for an acceptable method of temporary contraception, preferably a long lasting one, and secondly the support and back-up care offered to acceptors needs to be reinforced.

The system of laying down targets needs rationalisation. While the issue of recommending the abolishing of targets was debated, it was agreed that, in spite of its many disadvantages, it might be necessary to have some targets to ensure that a pressure for performance remained on the system. However it was clear that the present system of targets being laid down for individual staff was counter-productive in many ways. The final decision to opt for a surgical method of contraception is not a sudden insight prompted by a need for the incentive money. In most cases this step is the culmination of many little steps influenced by a host of interventions from the media, the health workers and from informal opinion seeking amongst family and peers. The fact that the 'case' is registered with a revenue or police officer at the penultimate moment is purely incidental and should not deprive the health worker of the credit for a long series of appropriate interventions.

We therefore recommended that the government consider the laying down of *area specific targets rather than personalised targets to be achieved by individuals.*

One problem clearly perceived relates to the linked issues of spacing and delaying the birth of the first child. It is clear that in the socio-cultural ambiance of our country it is unrealistic to expect that newly married couples can withstand family and social pressures and consciously delay having the first male child for even a short while.

The existing law regarding minimum age of marriage needs to be enforced strictly, and concerted efforts made not only for legal enforcement, but even more important, towards mobilising public opinion through intensive media efforts and social action. A few anomalies in the act need to be amended.

We feel that to push up the age of marriage it would be useful to generate a public debate on the desirability of taking a decision to give preference to unmarried men and women during selections to first level government and public sector jobs. It is not intended that only unmarried individuals be eligible for first level employment, but merely that, when all things are equal they should be given priority. As government service is still the ideal goal of a very large segment of our population from every walk of life, such a norm is likely to very rapidly and materially affect the preferred age for marriage in all social levels.

4   Medical Education

We are unanimous that the quality and focus of medical education needs to be changed to reflect national aspirations more faithfully. Though the Indian medical colleges have produced many outstanding and sensitive doctors, these are the exceptions rather than the rule. The system did not consciously set out to produce physicians who were equipped to handle the managerial requirements of doctors serving at the periphery, nor did it do enough to emphasise the socio-cultural and economic dimensions of health care.

To address this problem, the government should make efforts in two directions. Firstly, and as an immediate step, the teaching of preventive aspects, epidemiology, the social sciences, health economics, health education and management should be strengthened in all medical colleges in the country. At the same time, schools of public health need to be established, preferably in conjunction with existing medical colleges with postgraduate and undergraduate training functions.

Secondly, steps should be initiated to establish, on an experimental basis, alternate streams of medical education with a greater emphasis on problem solving for health care delivery, community level interventions and need based care.

We consider epidemiology, management and communication skills as areas of particular weakness in the medical education system. New entrants into peripheral health services often find themselves unable to cope with these aspects of their work even though they do not find the curative care needs taxing. It is therefore recommended that government establish a system similar to that for new entrants into the All India Services. This period of perhaps three months could be used to brush up the above mentioned areas and also to provide information on aspects such as service rules, financial rules, drug indenting, payment and disbursement of salaries etc..

The ROME Scheme was a laudable concept but, for various reasons, has achieved only limited success in India. It needs to be replaced by a broader based scheme taking the emphasis from the *ho* spital to the *pe* ople. This HOPE plan should ensure that each medical college is linked with a district health set-up. The government should not shirk ensuring that the medical college has direct responsibility and authority for the district health set up. Student instruction should be largely based on field and district hospital experience. The medical college could also be given an extension responsibility for another two contiguous districts for the purposes of emergency disaster consultation, epidemiological surveillance and local level planning. This would link all the districts in the country to one or the other tertiary care institution.

With about 147 medical colleges in India, already admitting some 15000 students, no new medical colleges should be opened. While this recommendation has featured both in the Planning Commission and Central Council documents, new colleges continue to be opened, even with government backing. Yet the conditions of many colleges is abysmal, with sadly deficient levels of competence, motivation and dedication in the staff, resulting in poorly prepared doctors more tuned to emulate their teachers in the business of medicine than in service to the community. There must be a moratorium on the opening of new medical colleges. Medical colleges who blatantly ignore MCI norms, faculty who are more concerned with a private practice are other examples of invidious ills besetting the education system.

The most expensive link in the health care delivery manpower structure is the doctor. Paradoxically, we are producing many times more doctors than nurses, inevitably delegating lower structured tasks to over qualified manpower. The imbalance must be corrected, not only passively by not allowing the further expansion of the base for training doctors, but by actively and selectively opening training facilities for nursing and auxiliary staff. There needs to be a special concentration of such institutions in the east, west and north. A rational manpower forecasting exercise needs to be done, with and production geared to meet these projected requirements.

The training of peripheral staff needs to be streamlined and augmented so that adequate staff is available to meet projected manpower needs. As institutions for training male multipurpose workers are largely supported by the states, there is a drastic shortage in several areas. In Rajasthan, for example, if all the MPW schools work to full capacity there will not be enough male MPW to man all the sub-centres until the next century. The training of auxiliary nurse mid-wives (ANM) is numerically better as this is a 100% centrally sponsored scheme under the family planning budget. However in many, if not most, instances the ANM training has a disproportionate emphasis on hospital based training and student ANMs are used in hospitals to fill in because of the severe shortage of nursing staff. The result is a worker poorly trained to do her prime duty as a field based female multipurpose worker, and motivated towards working as low grade nurse substitutes.

It is recommended that the opening of institutions for training male multipurpose workers be encouraged It is also recommended that the system of female multipurpose worker training be revised with a curriculum reflecting their changed role (from auxiliary nurse mid-wife to multipurpose worker), and including a mandatory field-based, instead of hospital-based, practical training.

Some form of continuing medical education is badly needed for all levels of health professionals. This can vary from post- induction training to orientation and refresher courses coupled with the provision for an ongoing stream of appropriate professional literature. In the course of our extensive collective field experience, we cannot recall any instance of professional journals being received by the PHC doctor. We do not remember ever having seen, or even heard of, written material for the para-professionals at the periphery. This is a serious lacuna for many reasons. Workers feel forgotten and neglected, and fall into a self perpetuating cycle of errors in their day to day work.

The institutions set up for this work are the health and family welfare training centres (HFWTC). It is the rare HFWTC that is effective and innovative in its work. Most of them function by rote and have no impact on the trainees. The doctors dislike going to them as the faculty is, in most cases, junior to them and in any case they have problems in collecting their TA and DA for the course. And yet recent experiences with the IEC Training Plan has shown that these very centres can be reactivated.

It is recommended that the HFWTCs, and state training institutions when established, be buttressed and invigorated and given the task of training all levels of health functionaries. In this effort they should draw upon the medical college associated with that district under the HOPE scheme. The centre should also ensure that the PHCs get professional literature to not only bolster the knowledge of the staff but also to remove their sense of professional isolation. It would also improve their quality by fostering an interest in academic medicine. Similarly material should be made available specifically for multipurpose workers so as to not only improve their knowledge but also infuse them with an *esprit de corps.*

The staff for the HFWTCs and state institutes could get their training in the schools of public health for the region.

We again reiterate that health care must be seen as a continuum ranging from care at the doorstep to the most advanced care at a tertiary care facility. Primary, secondary and tertiary care institutions are therefore equally important and what must be aimed at is the appropriate balance in the provision of facilities. The country must have tertiary care facilities of the very highest order, but we can only justify and afford this if adequate primary care is available to all. Training of manpower in the specialities and super-specialities must therefore continue, but along a pattern governed by national needs and manpower requirement forecasting.

5 Health Education and Community Participation

These two sectors are discussed together because health education is an important and integral part of the process of health empowerment of the people. There is little doubt that, in the final analysis, health cannot be delivered to the community. It is the people themselves who must lead the way and, change their behavior to a healthier pattern. The Governments role is therefore largely that of providing a basic service and infra-structure, and thus facilitating the adoption of a healthier lifestyle. It is also true that no government, specifically no developing country, can find the resources to met all the health and medical care needs of their citizens. Evolving high technology in diagnostics and medical interventions have further escalated the cost of medical care, and, as an inevitable corollary, further distanced the poor from the 'modern' hospitals.

The solutions are twofold. Government must give up the concepts of free medical care at all levels and the belief that it can deliver health to the people. The people must be encouraged to share the responsibility for their own health and medical care. It is not our intention to suggest that the government can wash its hands of the responsibility for the health sector; we are merely suggesting that the health care system must take an active role in empowering the people in matters relating to health. While there is every justification, and in fact need, for government to bear the major share of preventive and promotive health care, the people must share in the outlay on curative care. This can take the form of people paying token charges at the secondary care level and a more substantial partial responsibility for tertiary care costs. A system of health insurance, possibly at all levels, also needs to be explored.

Medicine has been mystified and kept out of the ambit of the consumer by both the doctor and the bureaucrat. It is now time that the burden is shared and the people encouraged to play an active role in looking after themselves and their families. Studies have shown that a large share of the total expenditure on health is being borne directly by the community itself, albeit not in payments to the official system. If adequate knowledge is made available to the community, the people can play an active role in helping themselves.

Health education cannot be imparted by doctors or the health para-professionals alone. Material for appropriate health education and information for transmission to the public must be devised jointly by the medical profession (who would provide the technical information) sociologists and anthropologists (who would translate it into a form acceptable by the people) and educationists (who are trained to communicate). The mere conveying of a technical message through the mass media or posters etc. has proved a dismal failure. Though there has been a dramatic change in the health messages through the mass media recently (and its impact has been clearly felt in the immunisation programme), the health departments education programme needs to be revitalised and modernised.

We recommend that health education, both through the national mass media, and using the local channels at the taluk, panchayat and even village level, be made a high priority activity of the health syst m. Student doctors during their education must get more emphasis on the socio-cultural determinants of health behaviour and be taught to develop communication skills.

One aspect of health education, though not in the conventional sense of the word, is the knowledge related to the services supposed to be available to the people. It is our impression that the vast majority of the people are not aware of the health and medical care services supposed to be available to them. This is especially true in the rural areas. Where experiments to inform the people of the arrangements and services have been tried, for example in the IEC Training Plan, this little measure has had considerable effect. The effect has been seen not only on the demand for, and utilisation of, services, but perhaps even more remarkably on the health system itself where workers and doctors found themselves accountable to the local community.

We recommend that every community, at least at the level of the sub-centre, and preferably at each village, be informed of the health services available to them locally.

As a part of the strengthening of the total health education network, special attention must be paid to the health education bureaus. Steps should also be taken to establish research in health education as an important and priority activity.

Steps should be taken to strengthen the social science inputs into the health system. This can be done at two levels. Firstly this aspect needs to be reinforced at the point of education of professionals and para-professionals, and secondly direct inputs need to be provided at the district, CHC and PHC level. Perhaps the district media and education officer (MEO) and the block extension educators (BEE) can be trained to take on this role. These two categories of staff are perhaps the most misused in the health set up and are used for almost every job other than health education inputs into the system.

447

6 Drug Policy

It has been estimated that some 45,000 medicinal drug formulations are marketed in India. At the same time bodies such as the WHO have suggested that some 258 drugs can meet essentially all the requirements of our country. The savings estimated from this step alone could be more than enough to meet the drug budget needs for all primary care institutions in the entire country. At today's prices it has been estimated that we need Rs 1000 crores to meet our drug needs. The expenditure today with the grossly inflated formulary is about Rs 2700 crores.

A definite policy decision needs to be taken to rationalise the formulary of drugs available in India today. While the figure of 258 drugs (and only 22 for primary health care) may be rather extreme, the available 45,000 is clearly an extreme in the other direction. In the interest of economy, rational patient care and good medicine it is important that policy level intervention is initiated.

7 Monitoring and the Health Information System

The unfortunate fact about the health information system is that in spite of the very intensive data collection effort at all levels, particularly at the peripheral level, almost no reliable statistics are available to those who can use this data in a meaningful manner. Amongst the reasons for this peculiar schism are the following:

● The long time between the gathering of data and its presentation in the final analysed form.

● The fact that disaggregate statistics are not available and national or even state statistics are meaningless in local level planning.

● The format of record keeping and for reporting is generally not designed with either health care or local level utilisation in mind, but is primarily an accounting system for generating numerical data on process events.

● As the staff collecting the data do not see any utility in that activity, the work is done carelessly and without interest, the entire exercise being treated as yet another imposition from above. This compromises the quality of data.

One example will suffice to illustrate this point. The Indian Council of Medical Research in 1989 reported that only 20.3 percent of the 198 PHCs they surveyed recorded 60% or more of births in their area, and none reached this level of achievement for recording infant deaths! (see *Fig. 3*)

Today there are about 15 registers and forms to be filled at the sub-centre level. This has been calculated by the Centre for Community Medicine to take an average of 843 hours a month. Thus 40% of the total working hours of the male and female multipurpose workers is spent in filling records and forms. This would have been bad enough even if something came of this effort. As the situation exists today, however, it is an activity that has almost no direct bearing on the health of the community, and certainly does not modify or change the health workers work plan, nor does it influence the PHC doctor except to enable him to see where he stands in the family planning sweepstakes.

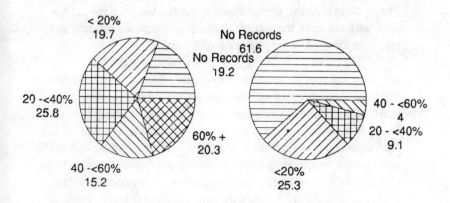

Fig. 3 Recording of births and infant deaths (source:ICMR)

449

The Central Council for Health and Family Welfare has resolved that the States initiate block level planning. We are in full agreement with this concept except that we recommend that, in an attempt to involve the people in their own health care, and to make the health care delivery system more ''of the people and for the people'', steps should be initiated to formalise block and district level planning. It must be ensured that people are involved, possibly through the panchayat system together with representation of organisations such as the Mahila Mandals. The great benefits of such community participation have been proved not only in many NGO projects but recently even in the government structure in the States of Bihar, UP, and Madhya Pradesh through the IEC Training Plan being supervised by the Centre for Community Medicine of the AIIMS.

Unfortunately block level planning is a mirage that is out of our reach today for several reasons. First and foremost is the fact that local level statistics are not available to aspiring planners either at the block or district level. Next is the problem that generally the leaders of the health team, either at the PHC or the district level, are not trained either to look critically at data or to plan interventions in the broad area of public health or even curative care. Another hindrance is the unfortunate exclusion of the people from the advisory or decision making process in the health field, specially at the peripheral level. The problem is further compounded by the fact that, in general, local staff does not have the flexibility to deviate from set patterns and priorities for work at the primary health centres.

All these barriers are amenable to solution. We are aware that the Ministry has been seized with the problem of a management information system for many years and has, in fact, been working towards evolving a suitable pattern for at least a decade or more. It is recommended that a new look be initiated at this problem, with the basic guidelines not being to try and include every thing possible, but to search for the minimum possible, consistent with good health care, evaluation and planning, both at the local and national levels. The data collected should have direct relevance to public health or managerial intervention, and the attempt should be not so much to satisfy the unreasonable demands of different programme officers, as to collect meaningful *minimum* information that is of relevance first to the periphery and then centrally.

The system of accounting rather than reporting meaningful statistics can be misleading. In immunisation, for example, the Ministry reported to the Central Council of Health and Family Welfare that the nation had achieved, during 1988-89, 109.2% and 101.2% of the proportional target for DPT and polio immunisation respectively. The earlier quoted ICMR study points out that only 29.2% and 23.2% of the PHCs had covered 80% or more of the children in their areas for the same vaccines. Neither figure is wrong in itself: the first is the accountant's view, the second is from the perspective of a health scientist!

In India there are several examples of experimental field records and a fresh look at the whole issue is long overdue. Some notable examples of cutting down both time and effort spent in record keeping have been developed by using microcomputers for first level data management at the PHC level. The examples range from those at the Bavla PHC initiated by Prof. Murthi of the IIM Ahmedabad, to that of the CMC at Vellore and the software developed by the Centre for Community Medicine (AIIMS) which is now operating at the Dayalpur PHC. The Dayalpur system has been evaluated and shown to reduce the total data recording and report generating time of a PHC (for a population of about 35000) from 849 to 315 hours a month. The time taken at the PHC level for generating all the reports needed for onward transmission was reduced from 27 hours and 50 minutes to 9 hours and 44 minutes. The computerised database system in the AIIMS example has been calculated to save Rs. 55,490 a year besides giving many other advantages, such as the generation of the reports needed by various agencies and also the monthly preparating of a work plan for each multipurpose worker detailing all the tasks to be done in each household of the area covered by the concerned sub-centre. The system is also responsive to queries made by the supervisors for forecasting possible trouble and planning future action. The system has been demonstrated to function, to save time and effort and thus free the workers for community service. It is this type of innovative approach that has helped this PHC to achieve many of our national health targets for the year 2000 AD. We recommend that the government explore the possibility of installing, on an experimental basis, some such system over a sufficiently large area and to link it with the National Informatics network.

The question of adequate training of medical manpower in the skills of critically looking at field data has already been discussed under the section on medical education. It is considered essential that the medical officer in charge of the community health centre and all district level staff be trained in public health and epidemiology. This approach has paid rich dividends where it has been implemented.

It is also recommended that a formal system of block and district level planning be introduced with the mandatory involvement of the local people through the Panchayat System and the Zilla Parishads.

A policy of permitting local freedom in deciding priorities and health interventions should be permitted as long as the national priorities are also included.

8   Summary and Recommendations

● The broad concepts of evolution of health care delivery infrastructure, as planned, are appropriate and can lead to the achievement of the general health goals of the country, if efficiently utilised.

● Several changes and mid-course corrections had become essential if the nation is to meet its broad objective of providing health care to all its peoples regardless of whether they were rich or poor, urban or rural.

● There is, by a large, enough knowledge, and adequate science and technology, to deal with most of the health care needs of our people.

● Adequate preventive, promotive or even curative care is not reaching the people, and the people themselves are generally not aware of the services already supposed to be available to them.

● Urban populations have very little provision for primary care, with the resultant overcrowding of tertiary care facilities. In contrast, facilities for appropriate referral are poor in rural areas.

● Both the medical profession and the bureaucrats have made no determined effort to de-mystify health and involve the people in their own health care.

● Though India was the first country in the world to officially sponsor family planning, even today the population *growth rate is high enough to be considered a crisis situation* which can and does nullify all our development efforts. There is a serious discrepancy between the couples stated to be protected and the actual birth rate.

● The system of monetary incentives is not producing the results anticipated.

● The currently available contraceptive technology does not meet the culturally influenced needs of service seekers. Intensive efforts need to be encouraged to develop more acceptable technologies. This must be in addition to continued efforts to stimulate increased acceptance of existing technologies.

● An adequate number of medical colleges have now been established to meet our general requirements. Efforts should now be concentrated on improving the quality and not the mere quantity of medical manpower.

● A serious distortion is evident in India in the manpower being produced for the health system. The *ratio of doctors to nurses and other paramedical professionals is irrational* and not conducive to efficient and cost effective health care delivery, neither is the distribution of medical specialities responsive to national needs.

● The drug policy in India is neither conducive to good health care, nor to cost effective deployment of resources.

- There appears to be very little, if any, interaction between medical colleges (who produce doctors) and the health system (who employs them).

- There is a serious lack of statistics, specially disaggregate data, and this hinders attempts to plan for smaller health care units (the available data only makes projections for the state or for the nation).

- The system of keeping records and reporting at the periphery leaves much to be desired. An undue amount of time is spent on this activity at the periphery, with no perceptible returns at the local level.

In view of the above, the committee made the following recommendations.

Health Care Delivery Infrastructure

The concept of having community health workers (CHW) is useful and a beneficial adjunct to the primary health care infrastructure. Its implementation, with greater emphasis on women's involvement, should be considered.

It is strongly recommended that *male multipurpose workers (MPW) be brought under one scale of pay* and the different designations, that are being kept alive only because of the different pay heads, be abolished. Some effective steps need to be taken to ensure that the manpower is actually appointed and then begins to work in tandem with the female worker. The actual MPW scheme envisaged functional changes upto the district level. Steps should be taken to implement the scheme in its entirety.

It is worth considering whether the Centre should take over the responsibility for the pay of not only the female MPW but also of the male worker. The existing pay slots thus freed for the state governments could be used for additional inputs into the rural health services.

We are in full agreement with the concept of the Primary Health Centre (PHC) and feel that this institution, which is planned as the backbone of the health care delivery infrastructure, needs to be strengthened and supported. While the population of 30,000 to be covered is small enough to permit intensive coverage, it is recom* mended that, as far as possible, a minimum of two doctors be posted at each PHC.

The concept of the community health centre (CHC) needs to be crystallised. This institution should take on the supervisory role for the four PHCs under its jurisdiction and also be responsible for the health information system and block level planning for the area. To do this it is essential that in addition to the four specialists already approved (pediatrics, medicine, obstetrics and surgery) steps should be taken to include a person trained in community medicine/ epidemiology.

Both the district media officer and the block extension officer are staff members who seem to do a variety of odd jobs. They are seldom used effectively for health education. These two categories of workers have important assignments which need emphasis, and should also serve as the social science inputs into the system.

It is recommended the the functions of the district level officers be modified so as to divide the PHCs and CHCs between them, each of them taking total responsibility for their area.

It must be emphasised, however, that unless an effective referral system is established from the most peripheral point to successively more evolved institutions, the credibility of the primary care system would remain dubious.

A system of urban primary health care must be developed so that not only do slum dwellers get easy access to health care, but tertiary care facilities avoid overcrowding and thereby are able to function as referral hospitals.

We strongly agree that national service is desirable. Even a two-year period of compulsory national service would make about 30,000 doctors available to the system. The requirement would be acceptable only if it was mandatory to all, regardless of any other considerations and if it applied to all professional courses, and not only to medicine.

Family Welfare Planning

Health and family welfare should be amalgamated into one comprehensive, integrated and inter-dependent activity (and department).

The system of monetary incentives for contraception needs to be abolished.

It is recommended that village development inputs be linked to family planning performance and not to population size.

Laparoscopic sterilisation should not be allowed at levels below that of a community health centre, and that too should be done by a specialist. At the PHC and similar levels the method of choice should be the mini-lap.

Efforts should be made to develop a culturally acceptable technology for temporary contraception.

It is recommended that the Government consider the laying down of area specific targets rather than personalised targets to be fulfilled by individual members of the staff.

The existing law regarding minimum age of marriage needs to be suitably amended and strictly enforced, and concerted efforts should be made not only for legal enforcement, but even more important, towards mobilising public opinion through intensive media efforts.

We feel that in order to push up the age at marriage, it would be useful to generate a public debate on the desirability of taking a decision to give preference to unmarried men and women during recruitment to first level government and public sector jobs. It is not intended that only unmarried individuals be eligible for first level employment, but merely that when all things are equal, they should be given preference.

Medical Education and Health Manpower

To correct the distortions in the medical education system, and to make the young doctors more in tune with national aspirations and priorities, the government should make efforts in two directions. Firstly, and as an immediate step, the teaching of preventive aspects, epidemiology, the social sciences, health economics, health education and management should be strengthened in all medical colleges in the country. At the same time schools of public health need to be established, preferably in conjunction with existing medical colleges, with postgraduate and undergraduate training functions.

Secondly steps should be initiated to establish on an experimental basis, alternate streams of medical education with a greater emphasis on problem solving for health care delivery, community level interventions and need based care.

Until such time that the medical education system in the country develops further, the government should consider starting a compulsory period of orientation and training for all new medical entrants into government service. These institutions could function along the lines of the pre-posting training of entrants into the All India Services.

The re-orientation of medical education (ROME) scheme was a laudable concept but for various reasons has achieved only limited success in India. It needs to be replaced by a broader based scheme shifting the emphasis from the *ho* spital to *pe* ople. This 'HOPE' plan should ensure that each medical college is linked with a district health set-up. The government should not shirk ensuring that the medical college has direct responsibility and authority for the district health set up. Student instruction should be largely based on field and district hospital experience. The medical college could also be given an extension responsibility for two more contiguous districts for the purposes of emergency/disaster consultation, epidemiological surveillance and local level planning.

There must be a moratorium on the opening of new medical colleges

457

any where in the country.

A rational manpower forecasting needs to be done, and production geared to meet projected requirements.

To correct the imbalance in medical manpower production in the country the government must establish training facilities for nursing and auxiliary staff. It is also recommended that the training of female multipurpose workers be revised with a curriculum reflecting their changed role (from auxiliary nurse midwife to multipurpose worker) and including a mandatory field-based, instead of hospital-based, practical training.

It is recommended that the opening of institutions for training male multipurpose workers be encouraged.

Some form of continuing medical education is badly needed for all levels of health professionals. This can vary from post- induction training to orientation and refresher courses coupled with an ongoing stream of appropriate professional literature.

It is recommended that the HFWTCs, and state training institutions, when established, be buttressed and invigorated and given the task of training all levels of health functionaries. In this effort, they should draw upon the medical college, associated with that district, under the HOPE scheme.

Health Education and Community Participation

The government must give up the concept of free medical care at all levels, and the belief that it can *deliver* health to the people. The people must be encouraged to share the responsibility for their own health and medical care.

While there is every justification, and in fact need, for government to bear the major share of preventive and promotive health care, the people must share in the outlay on curative care. This can take the form of people paying token charges at the secondary care level and more substantial partial responsibility for tertiary care costs. A system of health insurance, possibly at all levels, also needs to be explored.

The Health Departments education programme needs to be revitalised and modernised. We recommend that health education, both through the national mass media, and using the local channels at the taluk, panchayat and even village level, be made a high priority activity of the health system. Student doctors during their education must get more emphasis on the socio-cultural determinants of health behaviour and be taught to develop communication skills.

We recommend that every community, at least at the level of each sub-centre (preferably each village) be informed of the health services available to them locally.

As a part of the task of strengthening the total health education network, special attention must be paid to the health education bureaus. Steps should also be taken to establish research in health education as an important and priority activity.

Drug Policy

A definite policy decision needs to be taken to rationalise the formulary of drugs available in India today, starting with drugs purchased for the health care delivery system. The savings thus effected would not only permit a greater supply, but would actually improve medical care by making available a more rational choice of medicines for primary health care.

Monitoring and the Health Information System

It is urged that the system of records and forms at the periphery be rationalised and made less time intensive and yet, at the same time, more responsive to public health action. A microcomputer based system has been demonstrated to not only improve health care delivery but materially effect savings of money and time. We recommend that the government explore the possibility of installing, on an experimental basis, some such system over a sufficiently large area and link it with the national informatics network.

We also recommend that, in an attempt to involve the people in their own health care, and to make the health care delivery system more "of the people and for the people", steps should be initiated to formalise block and district level planning. It must be ensured that people are involved, possibly through the Panchayat and Zilla Parishad systems, together with representation of organisations such as the Mahila Mandals.

It is considered essential that all district level staff be trained in public health and epidemiology. Medical Officers incharge of the community health centres must also receive this training. This approach has paid rich dividends where it has been implemented.

A policy of permitting flexibility and local freedom in deciding priorities and health interventions should be authorised as long as the national priorities are not disturbed.

10
Fertilizer Use

With the present rate of growth of population of over 2 per cent per annum, India will need 225 million tonnes of food grains per year by 2000 AD. The fertilizer has been, and will continue to be, a key input in achieving this goal.

Contents

10
Fertilizer Use

Preamble

The fertilizer is the kingpin in India's Green Revolution and self-sufficiency in food. An increase in food production from 52 million tonnes (mt) in 1951-52 to 152.4 mt in 1983-84 has been possible with the introduction of high yielding rice and wheat varieties and a commensurate increase in fertilizer consumption from 65 thousand tonnes in 1951-52 to 8.7 mt in 1986-87. Although this success in food production has been commendable, there should be no room for complacency. With the present rate of growth of population of over 2% per annum, India will need 225 mt of food grains per year by 2000 AD; an additional production of 75 mt/year. The fertilizer has been, and will continue to be, a key input in achieving this goal.

Recognising the important role that fertilizers can play in increasing the food production in the country, the Government of India appointed a Committee on Fertilizers in 1985, headed by Shri B Shivaraman. The National Commission on Agriculture constituted in 1970 also went into the details of the plant nutrient needs for the projected food demands. The National Council of Applied Economic Research (NCAER) was asked by the Government to carry out a fertilizer demand study during 1975-77. A high powered committee on fertilizer consumer prices was constituted in 1985 under the Chairmanship of Dr G V K Rao. The main findings and recommendations of these important committees are discussed in the following pages.

The report of the Committee on Fertilizers was the first document to chalk out the course of development of fertilizer use as well as of the fertilizer industry in India. The report highlighted the critical importance of fertilizer-responsive varieties and the need for promoting a scientific and balanced use of fertilizers. It called for a special effort to promote the scientific use of fertilizers, adequate credit support both for production and distribution of fertilizers, maintaining reasonable prices, equalisation of price at railheads in the country and reasonable distribution margins to encourage efficient distribution. The Committee also recommended special measures for encouraging domestic fertilizer production to meet the rising demand. The Committee thus laid the policy foundation relating to fertilizer production, promotion, distribution and consumption.

The National Commission on Agriculture (NCA) in its report highlighted the importance of balanced fertilizer use on the basis of soil testing and technical advice given by the Departments of Agriculture, agricultural universities and other such organisations. It computed nutrient removals by crops, showed gaps between addition and removal of plant nutrients and made fertilizer projections based on the demands for food and other commodities. It pointed out that the wholesale and retail dealers must get a fair margin of profit to cover their costs. The margins must be such as to enthuse the cooperatives to reach the unprofitable interior areas which are often avoided by the private trade. It also suggested that fertilizer factories, their wholesale dealers, as well as retailers must do effective fertilizer promotion work. The NCA urged extensive use of fertilizers over a large area instead of intensive use in irrigated areas only. They also pointed out the likely possibility of micronutrient deficiencies due to cultivation of high yielding crop varieties and the use of high analysis fertilizers. The possible use of organic manures was suggested to partly rectify such a deficiency.

The National Council of Applied Economic Research (NCAER) was asked by the Government to make a fertilizer demand study in 1975-76. The study revealed that only 45% farmers in India used fertilizers and that only 33% of the total cropped area was fertilized. The majority of non-users were the cultivators with small holdings. The

NCAER study concluded that lack of irrigation was the major constraint followed by lack of credit. The irrigated area consumed 86% of the fertilizer. The NCAER study therefore suggested the expansion of irrigated area and distribution of improved varieties of seeds. The report also suggested that efforts should be made to develop the technology of fertilizer use in rainfed areas. The report pointed out that, among the economic factors, the relative price was the most important. Poor resources of small and marginal farmers and their lack of access to credit facilities were the major constraints.

The High-Powered Committee on Fertilizer Consumer Prices went into the details of promotion, distribution and marketing of fertilizers and suggested several measures. It recommended, and supported, the existing policies of uniform fertilizer price throughout the country. In order to bring stability in the market, the Committee recommended that the marketing territory of each fertilizer unit should be decided on a long term basis and frequent changes should not be allowed in the system. For hilly and rainfed areas, the Committee recommended small packages of 15 to 20 kg in place of the currently used 50 kg packages. The Committee pointed out that, at present, the retention price system, implemented by the Department of Fertilizer, does not include any provision for fertilizer promotion activity, with the result that manufacturers do not find any incentive for investing in fertilizer promotion. The Committee recommended that a provision for promotion activities should be allowed under the retention price system and highlighted the role of cooperatives. The Committee also looked into the agronomic factors responsible for efficient fertilizer use and laid emphasis on soil testing as a tool for efficient fertilizer use. It suggested a massive national project for promotion of fertilizer use in rainfed areas, and also noted the low efficiency of nitrogen applied to wetland rice. The role of recycling of organic wastes and biofertilizers was also highlighted. As regards the future product pattern, the Committee recommended the production of urea as a source of nitrogen, diammonium phosphate (DAP) as a source of P_2O_5 and muriate of potash as a source of potash. The Committee suggested that, in future, no capacity should be sanctioned for NPK complex fertilizers and granulated mixtures.

467

Historically the major issues in the country have been, and continue to be, production, pricing, marketing and distribution of fertilizers. The government, as well as the fertilizer industry, has been mainly concerned with the problem of low fertilizer use in India; on an average it was about 50 kg/ha of $N+P_2O_5+K_2O$ in 1986-87.

While a significant growth in fertilizer consumption has taken place, which has been at an average of 13.7% during the last 20 years, its impact on agricultural production has not been as perceptible as it should have been, particularly during the recent years. Fertilizer consumption increased during the period 1966-67 to 1980-81 by about 4.4 mt, whereas foodgrain production increased by about 46 mt, giving a response ratio of 12.5:1. However, from 1980-81 to 1985-86, an increase in fertilizer consumption of about 3.2 mt added 21 mt of foodgrains, giving a response ratio of 6.5:1 This shows that the efficiency of fertilizer use, as reflected by the response ratio, has, of late, decreased. In fact if we separate the additional food grains produced due to various other factors such as irrigation, high yielding varieties and residual fertility, the response ratio will be even more less. This indicates that the fertilizer is being used with very low efficiency. On the other hand, the reports from the agricultural experimental stations give a response ratio of 15:1 or more. In order to maximise agricultural production for the given fertilizer resources, it is, therefore, essential to increase its use efficiency. Taking these facts into consideration it was decided to constitute a Working Group for examining the situation, and for developing strategies for efficient and optimum fertilizer use in the country with the following terms of reference.

● To critically examine the kinds of fertilizers that will be best suited to Indian needs in relation to:

○ Food needs and the kinds of crops to be grown in different parts of India.
○ Use efficiency of major plant nutrients by different crops.
○ Soil test data from different regions of the country, and
○ Indigenous raw material availability in the country.

● To examine the different estimates of fertilizer needs of the country by 2000 AD and to arrive at realistic needs.

● To suggest new areas of research and study, with the overall objective of developing efficient fertilizer materials for saving energy and making the country self sufficient in fertilizers.

After its first meeting, held on 26 October 1987, the Working Group modified the terms of reference as under:

● Strategies for efficient and optimum fertilizer use.

   ○ Use efficiency of major plant nutrients by different crops, and available and potential technologies.
   ○ Improving soil test techniques and their use.
   ○ Packaging and marketing of fertilizer in relation to their efficiency and use optimisation.

● To critically examine the kinds of fertilizers that will be best suitable in relation to the kinds of crops to be grown in different parts of India.

● To examine the different estimates on fertilizer needs of the country by 2000 AD, and to arrive at realistic needs.

● To suggest new areas of research and development with the overall objective of developing efficient fertilizer materials, for saving energy, and the utilisation of indigenous raw materials for making the country self-sufficient in fertilizers.   ■

1   Fertilizer in Indian Agriculture Scenario

The production of chemical fertilizers in India started in 1906 with a single superphosphate plant in Ranipet. The next chemical fertilizer to be made in India was ammonium sulphate in Jamshedpur in 1919. Only about 10 per cent of the total production of about 4436 tonnes

Table 1
Statistics on food grain production and fertilizer use (5 yearly average) - beginning from 1st five year plan (All India)

Years	Gross cropped area (m.t)	Gross area under food grains (m.ha)	Total food grain production (m.t)	Total Fertilizer consumption $(N+P_2O_5+K_2O)$ (m.t)	Coverage under high yielding varieties (m.ha)	Gross irrigated area (m.ha)
1951-52 to 55-56	141.0	105.3	63.6	0.10	-	24.3
1956-57 to 60-61	150.5	113.4	74.0	0.23	-	26.9
1961-62 to 65-66	156.9	116.9	81.0	0.60	-	29.8
1966-67 to 70-71	161.7	121.0	94.2	1.73	8.8(7.3%)	35.3
1971-72 to 75-76	166.6	123.5	105.5	2.75	25.2 (20.4%)	40.6
1976.77 to 80-81	171.4	126.6	121.8	4.72	38.8 (30.7%)	47.4
1981-82 to 85-86	176.9*	127.8	142.3	7.42	51.5 (40.3%)	52.5*

* 3 yearly average from 81-82 to 83-84

Source: Fertilizer Statistics, 1986-87, Fertilizer Association of India.

of ammonium sulphate was used in India and the rest was exported. By 1925, 40% of the manufactured ammonium sulphate was consumed in the country. Later on, the chemical fertilizer became popular with the farmers and ammonium sulphate had to be imported. The Royal Commission on Agriculture in its report, submitted in 1928, emphasised the use of chemical fertilizers.

Five yearly averages of fertilizer consumption, food grain production and cropped area are given in Table 1. The average annual fertilizer consumption increased from 0.1 mt to 0.6 mt before the introduction of high yielding varieties (HYV) of cereals. After the introduction of the HYV programme, the average annual fertilizer consumption during the quinquennum 1966-67 to 1970-71 increased to 1.73 mt and by the quinquennum 1981-82 to 1985-86, it was 7.42 mt. Thus the per annum fertilizer consumption increased 74.2 times from the quiquennum 1951-52 through 1955-56 and 1981-82 through 1985-86. During the same period the increase in gross cropped area was 21.4%. The main factors responsible for an increase in fertilizer consumption were an increase in area under high yielding varieties and irrigation.

Nitrogen accounts for about two thirds, while phosphorus accounts for about one fourth, of the total NPK usage in India. The rest is accounted for by potash. Urea is the dominant nitrogen fertilizer contributing about 80% of the total N consumption. Diammonium phosphate and single superphosphate are the major phosphate fertilizers contributing about 50% and 16%, respectively towards the total P consumption. Muriate of potash is the major potassic fertilizer.

The progress in fertilizer consumption, in different zones, during the last 20 years is depicted in Table 2. During this period the fertilizer consumption went up 25 times in the northern zone, 7.5 times in the western, 9.5 times in the eastern and 5 times in the southern zones. Thus, most of the increase in fertilizer consumption has been in the northern zone and this was primarily due to the introduction and spread of dwarf wheats. A zonewise analysis shows that fertilizer consumption, as a proportion of total national consumption, is 43.9% in the northern zone, 24.4% in the southern zone, 13.6% in the eastern

471

zone and 8.2% in the western zone. However, it will be appropriate to mention that in 1983-84, the northern zone accounted for 22.1% of the total cropped area in India (180.4 million hectares). The percentage share of the total cropped area in India for the southern, eastern and western zones, in 1983-84, was 19.2, 18.1 and 40.6 respectively. The data of per hectare consumption of fertilizer show that, in 1986-87, Punjab recorded the highest consumption of 159.9 kg/ha nutrients followed by Tamil Nadu, which recorded 97.1 kg/ha. As against this the per hectare fertilizer consumption in Assam was only 4.7 kg, in Orissa 13.9 kg and, in Bihar, 41.4 kg. In the dry western zone the fertilizer consumption varied from 13.1 kg/ha in Rajasthan to 38.6 kg/ha in Gujarat.

An analysis of the districtwise consumption of fertilizers showed that, in 1985-86, 24 districts (9 in Punjab, 4 in Andhra Pradesh, 3 each in Tamil Nadu and Uttar Pradesh, 2 in Haryana, 1 each in Maharashtra, Karnataka and West Bengal) out of a total of 350 districts in 16 states consumed 25% of the fertilizer, while 69 districts consumed 50% of the fertilizer. A consumption of above 200 kg $N+P_2O_5+K_2$/ ha in 1985-86 only took place in three districts: Ludhiana, West Godavari and Nilgiris. In 122 districts, that is nearly one third of the total number of districts, fertilizer consumption was 25 kg $(N+P_2O_5+K_2O)$/ha or less, which is about one half of the national average (Table 3).

The cropwise analysis made by NCAER showed that rice and wheat crops used about 64% of the total fertilizer. Sugarcane was the next major consumer of fertilizer accounting for 8%, while cotton accounted for 5.5%. Consumption by groundnut was 2%. Commercial crops like potato, chilli, banana, tobacco, jute, plantation crops and other crops such as millets and pulses accounted for the rest of the fertilizer consumed.

An analysis of the fertilizer consumption pattern data in India shows that the fertilizer consumption has remained low in the following areas.

• Rainfed areas of the wet eastern zone.

Table 2
Zone and Statewise Trends in Fertilizer consumption ('000 tonnes)

Zone/State	1966-67	1971-72	1976-77	1981-82	1986-87
East Zone	*142.7*	*263.0*	*378.9*	*566.0*	*1255.0(39.5)*
Assam	5.5(1.9)*	8.2(3.0)	3.8(1.2)	10.8(3.3)*	16.8(4.6)
Bihar	64.7(7.0)*	108.3(9.8)	155.9(13.8)	205.3(18.0)*	527.5(51.2)
Orissa	23.9(3.2)	49.0(5.8)	62.0(8.0)	82.0(9.9)	15.8(17.3)
West Bengal	48.6(7.3)	95.4(13.5)	152.5(19.2)	258.4(32.8)*	499.2(65.8)
Manipur	-	0.8(4.3)	1.98(9.4)*	3.3(15.4)*	5.66(30.4)
Meghalaya	-	-	1.85(9.1)*	2.2(9.5)*	3.48(16.1)
Nagaland	-	-	0.12(1.1)*	0.35(2.0)	0.35(1.8)
Sikkim	-	-	-	0.79	1.36(10.9)
Tripura	-	1.27(3.8)*	0.78(0.4)*	2.68(7.6)*	6.84(19.3)
Arunachal Pradesh	-	-	-	0.10	0.33(1.8)*
Mizoram	-	-	-	0.049	0.15(2.4)*
North Zone	*175.3*	*849.3*	*1265.5*	*2391.4*	*3370.2(84.7)**
Haryana	-	82.1(16.6)	137.3(25.2)	251.6(45.5)	414.9(75.3)*
Himachal Predesh	0.4	8.0(8.8)	9.1(9.8)	18.2(19.5)	26.1(26.3)*

Table 2 (Contd.)

Zone/State	1966-67	1971-72	1976-77	1981-82	1986-87
Jammu & Kashmir	3.3(4.1)	5.7(6.6)	12.4(13.5)	21.6(21.8)	30.3(29.5)
Punjab	66.6(12.8)	289.5(52.6)	371.0(59.3)	820.5(123.7)	1115.6(158.5)
Uttar Pradesh	104.8(4.7)	463.0(20.2)	729.4(31.4)	1269.6(52.2)	1771.7(70.5)*
Chandigarh	-	-	-	1.0	1.07(269.5)
Delhi	-	0.96(8.6)	6.3(52.9)*	8.9(74.9)	10.50(114.0)
South Zone	*478.6*	*881.2*	*960.3(28.0)*	*1659.6(47.0)*	*2309.6(68.1)*
Andhra Pradesh	201.1(15.9)	297.0(22.6)	401.5(31.1)	655.5(50.0)	901.5(73.9)
Karnataka	74.4(7.1)	166.8(15.5)	206.2(18.5)	383.6(34.4)	565.8(48.5)*
Kerala	54.3(20.7)	65.0(22.3)	69.3(23.2)	94.8(32.9)	151.4(52.7)*
Tamil Nadu	148.8(20.4)	346.0(48.3)	277.6(38.4)*	512.6(66.7)	674.5(95.2)*
Pondicherry	-	6.4(125.5)*	5.7(105.7)*	13.1(2555.8)	16.3(338.8)
A & N Island	-	-	-	0.007	0.14(4.7)
West Zone	*238.3*	*607.6*	*731.2*	*1309.4*	*1803.7(25.6)*
Gujarat	65.1(6.4)*	182.4(17.9)*	202.4(19.9)	401.4(38.6)	402.3(39.2)
Madhya Pradesh	48.1(2.6)	123.0(6.0)*	136.6(6.3)	236.3(10.9)	493.6(22.1)

Table 2 (Contd.)

Zone/State	1966-67	1971-72	1976-77	1981-82	1986-87
Maharashtra	106.(5.6)	241.4(12.4)*	290.0(14.7)	529.1(26.6)	655.0(32.1)
Rajasthan	18.6(1.2)	58.5(4.1)*	98.7(5.7)*	138.0(7.9)*	247.1(14.3)
Goa Daman & Div.	-	2.3(16.5)*	3.5(25.2)*	4.4(30.3)	4.4(28.0)
Dadra & Nagar Haweli	-	-	-	0.23	0.43(16.6)
All India	1100.6(7.0)	2656.9(16.2)*	3411.0(20.0)	60.67.2(34.6)	87.38.4(49.4)

* (Kg/ha)

Note: Aggregate of zones does not tally with all India total as the latter includes plantations for 1966-67, 1971-72, 1976.77 and 1981-82.

Table 3

Classification of districts according to ranges of fertilizer consumption (kg/ha) of gross cropped area - 1985-1986

State	Above 200	150-200	100-150	75-100	50-75	25-50	10-25	5-10	Less than 5	No. of dists.
Andhra Pradesh	1	3	1	-	6	10	2	-	-	23
Assam	-	-	-	-	-	-	-	4	6	10
Bihar	-	-	1	7	7	9	8	1	-	33
Gujarat	-	-	1	2	2	8	4	1	1	19
Haryana	-	-	3	2	3	3	-	1	-	12
Himachal Pradesh	-	-	1	-	-	4	5	2	-	12
Karnataka	-	-	3	2	6	6	3	-	-	19
Kerala	-	-	-	2	5	6	-	-	-	13
Madhya Pradesh	-	-	-	-	1	13	20	5	6	45
Maharashtra	-	-	1	-	3	13	7	1	-	25
Orissa	-	-	-	-	-	2	6	3	2	13
Punjab	1	8	2	1	-	-	-	-	-	12
Rajasthan	-	-	-	-	-	3	8	7	8	26
Tamil Nadu	1	1	6	5	1	3	-	-	-	17
Uttar Pradesh	-	1	13	15	13	5	4	4	1	56
West Bengal	-	-	2	1	5	6	1	-	-	15
Total (16 states)	3	13	34	37	52	89	68	30	24	350

Source : Fertilizer statistics (1986-87).

● The dry western zone having larger areas under pulses and oilseeds. In addition, there are areas under problem soils such as acid, saline/alkaline and acid sulphate soils. In these areas also, fertilizer consumption and its efficiency is low.

2 Strategies for Efficient and Optimum Fertilizer Use

2.1 Use Efficiency of Major Plant Nutrients

The agronomists, soil scientists and plant physiologists have different concepts of fertilizer use efficiency (FUE). The agronomists and economists use FUE for the expression 'kilogram of produce per unit fertilizer or output per unit input, while soil scientists prefer true or apparent recovery of the applied nutrient. Plant physiologists use the term production efficiency of a nutrient, which is increase in yield per unit of nutrient absorbed. However, at the farmer's level, and for price planning, the increase in yield of economic produce per unit of applied fertilizer is the relevant criterion. This is also referred to as the 'yardstick' when used for calculating the additional production due to fertilizers. Yardsticks of additional production due to fertilizers, for some crops, are given in Table 4. These yardsticks were calculated from the data of a large number of trials on the cultivator's fields under the All India Coordinated Agronomic Research Project (AICARP) of ICAR. Some important points brought out by the data in Table 4 are:

● In irrigated agriculture, FUE of nitrogen was much higher in wheat than in rice.

● FUE of nitrogen in rainfed rice was very low.

● The response of rabi rice to P and K was as good as that to N. There is, thus, considerable scope for balanced fertilization.

● Of the coarse cereals, grown under rainfed conditions, maize responded to nitrogen better than sorghum.

Table 4
Yardsticks of additional production from fertilizer (kg grain/kg nutrient) - Data from AICARP (1977-82)

A) Irrigated

	N 40*	N 80	N 120	P 40 over N 80	P 60 over N 120	K 40 over N 80 P 40	K 60 over N 120 P 60
Rice (kharif)	12.7	11.4	10.8	10.5	8.8	8.3	7.2
	(8.5)**	(7.6)	(7.2)	(7.0)	(5.9)	(5.5)	(4.8)
Rice (rabi)	9.8	8.6	7.8	9.9	8.0	9.2	8.0
	(6.5)	(5.7)	(5.2)	(6.6)	(5.3)	(6.1)	(5.3)
Wheat	13.7	11.4	10.4	9.5	8.2	6.0	7.5

B) Rainfed

	N 30	N 60		P 30 over N 60	P 60 over N 60	K 30 over N 60	P 60
Rice (kharif)	6.1	6.0		11.0	8.5	6.2	
	(4.1)	(4.0)		(7.3)	(5.7)	(4.2)	
Wheat	8.3	7.3		7.7	6.4	7.5	
Maize (kharif)	10.8	8.5		8.4	6.1	6.7	

478

Table 4 (Contd.)

	N 30	N 60	P 30 over N 60	P 60 over N 60	K 30 over N 60	P 60
Sorghum (kharif)	7.4	6.7	6.8	5.2	7.0	
Sorghum (rabi)	5.1	4.7	6.1	4.6	5.3	

	N 20	P 20	P 20 over N 20	P 40 over N 20	P 40 over N 20	K 20 over N 20	P 40
Black gram (kharif)	7.2	4.5	6.3	4.0	5.2	-	2.8
Black gram (rabi)	17.2	5.4	7.4	4.9	6.0	-	4.3
Chickpea (rabi)	15.4	8.9	8.7	7.2	7.6	-	3.8
Pigeonpea (kharif)	14.2	7.2	8.5	6.0	7.0	-	2.1

* Kg N/P$_2$O$_5$/K$_2$O/ha

**Figures in parenthesis are for rice grain after shelling (Taken as two third of paddy).

Source: Randhawa, NS, Bhargava, P.N. and Jain, H.C. (1985). Current status of yardsticks of crop response to fertilizer. Proc. FAI Group Discussion on Means to Increase Crop Response to Fertilizer Use, New Delhi, Sept. 4-5, 1984.

● Maize and sorghum also responded well to P and K. (since the data are for rainfed conditions, these bring out the need for balanced fertilization even under rainfed conditions).

● Pulses give a good response to starter N (a small amount given at sowing).

● Pulses give a good response to P. It may be desirable to take a look at the data on recovery of applied plant nutrients. Data on recovery of fertilizer N (Table 5a) show that only about one third or even less, of the applied urea is recovered in rice crop. Recent N studies have shown (Table 5b) that about 31-35% of applied urea was recovered by the rice crop, 3-4.6% by the succeeding wheat crop and 16-25.6% was retained in the oil. About 56 to 60.6% of the applied N could not be accounted for and was lost.

Data on the recovery of applied P are given in Table 5c. These data show that the cereals recovered about 12 to 29% of applied P, while

Table 5a

Recovery of fertilizer N by kharif cereals

Crop	Source of N	N rate (kg N/ha)	N recovery* (%)
Rice	Urea	100-200	36.0
	Urea briquettes	100-200	44.0
	Urea coated with neem cake	100-200	41.5
	Sulphur coated urea	100-200	40.0
	Urea		18-29**
Maize	Urea	60-180	63.1-40.3
Sorghum	Urea	120	58.5
Wheat	Urea	100-200	40-90

* Apparent recovery determined by the difference method,
** 15N data.
Source: Prasad, R. and Subbiah, B.V. (1982). Fertilizer News 27(2):27-112

legumes recovered only 6.4 to 18.4% of applied P. The rest of the applied P remained in the soil and only small amounts became available to the succeeding crops. Appropriate strategies have, therefore, to be developed for increasing the efficiency of both nitrogenous and phosphate fertilizers.

2.2 Rainfed Areas of Wet Eastern Zone

The fertilizer consumption in the wet Eastern zone (Table 2) is low as the major crop is rice, where, again, the efficiency of applied fertilizer nitrogen is low (Tables 5a & 5b).

Table 5b

15N balance in rice-wheat double cropping (percentage of applied N)

Kg N/ha applied to rice	Recovered by				Leaf in soil		Total accounted N	
	Rice		Wheat					
	Summer fallow	Green manuring	Summer fallow	Green manuring	Summer fallow	Green manuring	Summer fallow	Green manuring
60	35.4	32.8	4.1	3.0	16.7	22.6	56.2	58.4
120	31.2	33.2	4.6	3.1	25.6	24.3	61.4	60.6

Source : Goswami, N.N., Prasad, R, Sarkar, M.C. and Singh, S. 1988. J.Agric.Sci. Cambridge (in press).

Table 5c

P utilisation by cereals and legumes (in per cent)

Legumes		Cereal	
Soybean (Var. Bragg)	8.2	Rice-221	29.1
Cowpea (C-152)	18.6	Maize (Ganga 5)	19.1
Greengram (PS-16)	6.4	Barley (DL-36)	14.8
		Wheat (Sonalika)	12.7

The major problem of the area is water management, particularly in low lands, where periodic drainage can considerably increase the efficiency of applied fertilizer. Data from a study at the Central Rice Research Institute, Cuttack (Table 6) clearly bring out the role of drainage in low land rice cultivation. Data on a scientifically designed drainage system are available from tea plantation and are given in Appendix-1. The provision of adequate drainage in lowland rice areas can go a long way in increasing the rice yields. It will also take care of the iron toxicity in these areas.

A detailed account of rice cultivation in Eastern India is given in Appendix-2. Top dressing of nitrogen is not feasible in the rainfed low lands growing rice. These areas would need fertilizers that can stay in the soil for a long time (such as slow-release N fertilizers) so that, when applied at sowing/trans-planting, nitrogen is available through-out the growth period of rice. Nitrification inhibitors can also increase the efficiency of applied urea. The available technologies include:

Table 6

Effect of frequency and depth of drainage on IR-8 Paddy (Mudholkar and Pradhan, Cuttack, 1969)

Depth of free water	Yield (q/ha)	Drainage frequency	Yield (q/ha)
Land submergence	51.5	Once between CTPF*	61.4
At soil surface	57.1	Twice between CTPF*	63.4
At 15 cm soil depth	60.0	Weekly	59.8
At 30 cm soil depth	62.5	* Period between critical tillering and primordia formation.	NS
At 45 cm soil depth	66.5		
CD 5%	5.3		

Source : Dastane, N.G.; Singh, M., Hukkeri, S.B. and Vamadevan, V.K. (1970). Review of Work Done on Water Requirements of crops in India. Navabharat Prakashan, Poona-28.

incubation of urea with moist soil for 24 hours before application, making of fertilizer mudballs and their use and using urea coated with neem cake.

Adequate P and K fertilization is necessary to increase the efficiency of applied N. Therefore, for higher fertilizer efficiency, balanced fertilization of NPK is required. An available low cost technology is to dip the roots of rice seedings in a slurry of phosphate fertilizer before the seedlings are transplanted. This technology can also be used for Zn where the zinc deficiency is reported.

The full potential and contribution of biofertilizers, specially blue green algae (BGA) and *Azolla,* are yet to be exploited. These can play an important role in this region. The necessary infrastructure for making the cultures of these organisms available to the farmers needs to be provided.

Organic manures (Appendix-3), specially green manures, (Table 5b) can help in economising on fertilizer and will help in recuperating soil fertility.

2.3   Dry Western Zone

The fertilizer consumption in the western zone is the lowest in the country (Table 2), while the cultivated area is the largest in this zone. Mostly dryland agriculture is practised in this zone.

The special characteristics of the dryland agriculture are:

● Low (less than 750 mm) and uncertain precipitation.

● Mostly short duration course cereals (sorghum, pearl millet), pulses (green gram, black gram, pigeon pea, kidney bean etc.) and oilseeds (sesame, groundnut, castor etc.) are possible.

● When the rainfall is good, well-distributed, and a sufficient stored soil moisture is available, double cropping is possible although most farmers skip the kharif crop and grow chick pea, mustard or sunflower during the *rabi* season.

● The short growing season forces the farmer to concentrate on seed procurement and sowing.

● A long dry spell, immediately after sowing, forces the farmer to go for re-sowing.

● In crops like bajra, having a very weak plumule, a heavy shower 3-4 days after sowing (which is frequent) leads to crust formation and seedlings are not able to emerge. Re-sowing thus becomes necessary.

● A mid-season long dry spell may completely dry out the crop, particularly when the soil moisture storage is low.

● Early cessation of the monsoon leads to poor grain filling and in many cases only fodder is harvested.

● Farmers generally follow mixed cropping - sowing mixed seeds of two to four crops to safeguard against total crop failure.

● Most of the fertilizers used in dryland areas is confined to cash crops like cotton, groundnut etc.

There is a wide variability in the types of soils in dryland areas, with their different soil depth and moisture holding capacity (Table 7). Suitable crops and cropping systems have been developed for different soil-depth stored moisture situations. This knowledge has to be extended to the farmers and adequate seeds and fertilizers have to be made available.

A good response to moderate levels of nitrogen is obtained (Table 8a). On alffisols (red soils) and entisols, a good response to phosphate is obtained. The response to phosphate on vertisols (black soils) is not obvious due to the high P fixing capacity of these soils.

The placement of phosphate is an important technology, available for increasing the efficiency of fertilizers in dryland agriculture. With the exception of wetland rice and fertilizers having low/no water soluble P, it is a proven practice, and widely recommended, both for irrigated and rainfed farming systems. Farmers can also deep place the initial N and K to get the maximum advantage. It was found that the drilling of fertilizer P brought about additional yield advantages of 2-3 q grain/ha, as a result of its better utilization by the crop, compared to the same dose applied as surface broadcast. It is obvious that the yield increases are adequate to pay for all or most of the investment in P applied. These results also show that P fertilizers can be costly for a farmer who adopts poor application techniques, but a paying proposition for another farmer who appreciates the value of proper P application.

Towards promoting efficient use of P, the large scale production of suitable fertilizer application machines is a worthwhile sector for the consideration of the industry.

Biofertilizers, specially *Rhizobium* cultures, could play an important role in increasing the efficiency of fertilizers applied to pulse crops. The necessary infrastructure, for making the *Rhizobium* culture available to the farmers, needs to be provided.

Organic manures can play an important role not only in meeting a part of the plant nutrient needs, but their application, along with chemical fertilizers, can also increase the efficiency of the latter. Organic manures also increase the water holding capacity of soils and this is of special significance in dryland agriculture. It is estimated that, if all the available resources are mobilised, organic manures have an annual capacity to supply about 8 mt of N, 2.66 mt of P_2O_5 and 4.3 mt of K_2O (Appendix-3). However, operational feasibilities and the economics of mobilising these resources need to be carefully studied. Research on speedy preparation of compost, utilising appropriate micro-organisms, also needs to be encouraged.

The point that is to be brought home to the farmers in dryland regions is that fertilizer application permits better utilisation of available soil moisture (Table 8b). Thus for better fertilizer use-efficiency, an integrated management of fertilizers and moisture is a must.

Table 7

Broad soil groups and their moisture storage capacities in rainfed areas of India

Sl. No.	Broad soil group	Sub-group (based on soil depth)		Moisture storage capacity (mm)	Suggested cropping systems
1.	Black	1.1	Shallow to medium (upto 45 cm)	135-145/45 cm	Single cropping
		1.2	Medium to deep (45-90 cm)	145-270/90 cm	Inter cropping
		1.3	Deep (90 cm)	300/m	Double cropping
2.	Red	2.1	Shallow to medium (upto 45 cm)	a) 40-70/45 cm (sandy loam)	Single kharif cropping
				b) 70-100/45 cm (loam)	Inter cropping
		2.2	Deep (>90 cm)	c) 180-200/90 cm	Double cropping
3.	Sierozem	3.1	Medium to deep (upto 90 cm)	80-90/90 cm	Single cropping

Table 7 (*Contd.*)

Sl. No.	Broad soil group	Sub-group (based on soil depth)	Moisture storage capacity (mm)	Suggested cropping systems
4.	Sub-montane	4.1 Deep	a) 90-100/m (loamy sand)	Inter cropping
			b) 110-140/m (sandy loam)	Double cropping
			c) 140-180/m (loam)	Double cropping
5.	Alluvial	5.1 Deep	a) 110-140/m (sandy loam)	Double cropping
			b) 140-180/m (loam)	Double cropping

Source : Randhawa, N.S. Singh, R.P. Fert. News. 28(9):17-32 (1983).

In addition to the climate and soil problems mentioned earlier, there are technological lacunae and socio-economic problems in drylands. The technological lacunae include the lack of high fertilizer responsive varieties, particularly of pulses and oilseeds, the pest and diseases susceptibility of available crop varieties, remunerative cropping systems, ineffective weed control and the lack of appropriate and viable resource management technologies. The socio-economic problems include the lack of credit to buy fertilizer and other inputs, and

Table 8a
Response of rabi sorghum to nitrogen

Centre	Rainfall (mm)	Kg N/ha	Grain yield (g/ha)	Response kg grain/kg N
Bijapur	680	30	16.6	9.7
Solapur	722	30	14.5	19.0
Bellary	500	40	15.8	6.5
Kovilapatti	730	30	23.2	27.7

Source: Venkateswarulu, J. Proc. FAI Group Discussion on Fertilizer use in Drylands. 3 (1979).

Table 8b
Moisture use efficiency (MUE) in pearl millet as influenced by fertilizer application

Treatment	MUE (Kg grain/mm-ha of moisture)				
	1975	1976	1977	1978	Average
Control (no fertilizer)	8.5	2.3	5.0	2.9	4.7
20 kg N/ha every year	10.4	5.0	7.5	3.3	6.5
40 kg N/ha once in 2 years	11.1	4.4	8.4	3.3	6.8
40 kg N/ha every year	12.8	6.6	8.8	4.9	8.3

Source : Singh, R.P., Singh, H.P., Daulay, H.S. and Singh K.C., Indian J. Agric. Sci. 51(7):498 (1981).

absences of a procurement policy for pulses and oilseeds, the major cash crops of the region.

A close linkage between the concerned technical, financial and the administrative machinery and the development of dryland agriculture is a must. This can be best achieved by creating an organisation on the lines of the Command Area Development Authority.

The adoption of complementary land use system like agro-forestry and animal husbandry, horticulture, sericulture and poultry development programmes would improve the farmer's financial position and his capacity to invest in fertilizers and other inputs for crop production.

2.4   Problem Soils

The strategy for increasing the efficiency of fertilizers in problem soils has to be, first, to remedy the soil malady (whether it is salts, alkalinity or acidity).

Data on the effect of application of lime on acid soil are given in Table 9a. The benefits of fertilizer application improved considerably when lime was also applied.

Table 9a

Yield of some crops (q/ha) as affected by liming and NPK in acid soils

Crops	Treatment			NPK + Lime
	Control	Lime	NPK	
Wheat (average of 5 years)	6.5	8.3	14.7	17.6
Barley	7.1	7.3	18.9	23.3
Maize	2.7	5.3	11.2	17.2
Jute	17.2	23.3	26.3	32.8

Source: Mandal, S.C., Sinha, M.K., and Sinha, H. (1975). Tech. Bull. (Agric.) No.51, ICAR, New Delhi.

Similarly, on sodic (alkaline) soils, gypsum application (with adequate fertilizer) can considerably increases the crop yield (Table 9b).

2.5   Soil Testing

In India, more than 450 soil testing laboratories, including 80 mobile soil testing laboratories, are functioning at present. There are about 89.4 million farm holdings out of which 12.4 million are irrigated and 16.9 million are partially irrigated. The analytical capacity of the existing laboratories is about 6 million soil samples per annum. Even to analyse one typical field in each holding once in 3 years, we will need about 30 million samples per annum.

The Dr Rao Committee report mentions that a survey conducted on the extent of collection of samples showed that 93.26% samples were

Table 9b

Effect of gypsum application on crop yields (q/ha) in sodic soil

Gypsum (t/ha)			1972-73		1973-74	
1971	1972	1973	Rice	Wheat	Rice	Wheat
6.5	-	-	70.9	40.5	64.1	50.6
6.5	6.5	-	70.9	39.1	62.2	47.2
6.5	6.5	6.5	-	-	67.1	48.5
13.0	-	-	68.0	45.8	58.9	49.4
13.0	6.5	-	71.0	41.9	67.4	49.3
13.0	6.5	6.5	-	-	64.6	51.6
19.5	-	-	67.9	44.6	62.9	50.2
19.5	6.5	-	65.8	47.0	70.1	50.3
19.5	6.5	6.5	-	-	63.8	51.7
26.0	-	-	65.8	44.8	65.8	51.2
26.0	6.5	-	69.1	48.8	64.9	51.0
26.0	8.6	6.5	-	-	63.9	53.6

Source : Central Soil Salinity Research Institute, Karnal, Annual Reports, 1971-79.

collected by officials of the Agriculture Department, while only 6.74% samples were brought by the farmers directly. This sums up how well-accepted the soil testing reports are and calls for a closer look at the quality of soil testing as a tool for fertilizer recommendation in India.

Considering the wide variability in soils, climate and crops, much more research is needed to improve the soil testing technique. Another factor which must be considered is that the classification of low, medium and high nutrient status is general, and not based on individual crops. For instance, an area classified low in P, may be so for wheat but would be medium for lowland rice. Similarly the critical level to classify the soil into various availability classes may be widely different depending upon the soil texture, mineralogy and soil order such as alfisol or vertisol. Data on these aspects are available in the literature for quite a long time now, but these have not yet been taken into account. There is, however, an urgent need to redefine the classification into low, medium and high categories with respect to all major nutrients and Zn. This should be done at the state level by local experts.

On the basis of the soil test techniques available, a mass of data has been generated during the last twenty years or so on making fertilizer prescriptions for different yield targets. These have, unfortunately, found limited acceptance and even more limited adoption. The entire effort of the soil test and crop response coordinated project of the ICAR is directed towards developing crop-wise fertilizer recommendations based on yield targets. However, fertilizers alone do not determine the yield. Crop yield is a complex expression of a number of factors such as genetic potential, environment, availability of plant nutrients and moisture, agronomic practices and the plant protection measures adopted. It is essential that soil test crop response correlation studies be conducted, on a number of farmers' fields, for a given bench mark soil, within the framework of the most popular cropping system. A systematic collection of information from these trials, over a number of years, may help to evolve a simulation model, which can be used for a generalised fertilizer recommendation in that region.

Furthermore, most soil tests have been developed based on data from irrigated experiments. Oilseeds and pulses, which are grown mostly under dryland conditions, have therefore received virtually no attention from the scientists responsible for soil testing. Most pulses have deeper root systems that can forage larger volumes of soil and, therefore, can feed better on native plant nutrients. The relationship between the response of these crops and the fertilizer needs to be carefully determined.

The information on the use of soil testing is currently available only on a single crop basis. The emphasis should be shifted towards generating results for different cropping systems as a whole, in various agroclimatic regions of the country, and also for different levels of investment, to cater to farmers with different resources.

A widespread deficiency of zinc and sulphur has been reported in certain regions. Very few soil testing laboratories can take up the analytical work for secondary and micro nutrients. Soil testing laboratories have, therefore, to be equipped to cater to need.

There is a need to convert soil fertility research into a vigorous discipline devoted to understanding soil-fertilizer relationships, (with an eye on furnishing dependable and adoptable advice to the farming community, which is currently spending Rs. 4300 crores annually on fertilizers). The system of developing soil fertility maps is, at present, both weak and outdated especially for classifying soils into different fertility categories. This has often led to results which defy interpretation. Soil tests have, therefore, come in for severe criticism in recent years. This criticism can only be countered if the 'mechanical meaning' given to soil tests is discarded, and the 'chemical meaning' takes its due place. Soil test research should be reorganised into a soil fertility evaluation research, and not wedded to any philosophy other than providing technical back-up for soil testing services, and providing deeper insights into the availability of the fertilizer to the crop.

Plant analysis as a diagnostic tool is more useful for horticultural and plantation crops, rather than annual crops. To direct the course of efficient fertilizer use, soil testing laboratories in areas which use moderate to high amounts of fertilizers, or where the horticultural crops are important, may be upgraded to facilities for soil and plant analysis.

It is time that the entire soil testing programme, both at the STCR and state levels, is thoroughly modified to provide a dependable service to the farmers. An integration of soil testing with soil taxonomy is required to provide fertilizer recommendations based on soil types. It is also necessary that the All India Micronutrient Coordinated Project develops coordination linkages with the Soil Test Crop Response Project.

2.6 Packaging, Distribution and Marketing of Fertilizer in Relation to its Efficiency and use Optimisation

Each fertilizer manufacturer has his own concept of an economic marketing area, depending on the objectives and concepts of his organisation. The general tendency is to increase the number of outlets in the profitable zone in order to minimise costs: it is the effort to increase consumption ('sales') which receives top priority; the effort to increase the fertilizer efficiency, unfortunately, receives the lowest priority in the commercial organisation's scheme of things!

The aim of an efficient fertilizer distribution network should be to see that the farmer does not travel more than 5-8 kms to get his requirement of fertilizers. In the case of low potential areas, the limits could be about 10-15 kms. We have to bear in mind that there are 6 lakh villages in India and nearly 40% of these hardly use any fertilizer (Chandra,1987). The available information (Table 10) shows that in the states of Assam, Bihar, Himachal Pradesh, Madhya Pradesh, Rajasthan and Uttar Pradesh too many villages are served by a fertilizer sale point. Obviously, during the kharif season, a farmer has to take considerable risk in transporting fertilizer from sale point to his farm due to rains. Heavy rains can lead to severe loss of the fertilizer.

A study conducted by NCAER during 1975-76 revealed that only 35 to 45% of the small farmers (less than 2ha; 25% of the area is held by

493

Table 10

Number of villages, gross cropped area and fertilizer consumption per sale point in different states of India - 1986-87

State	No. of villages	No. of* fertilizer sale points	No. of villages served per sale point	Gross cropped area covered per sale point (ha)	Fertilizer consumption per sale point (tons)
Andhra Pradesh	29,428	9,778	3	1370	92.2
Assam	23,331	3,276	7	1092	5.1
Bihar	78,027	8,840	10	1161	59.7
Gujarat	18,697	8,097	2	1287	49.7
Haryana	7,064	3,853	2	1476	107.6
Himachal Pradesh	18,929	2,350	9	413	11.1
Jammu & Kashmir	6,742	1,495	5	676	20.3
Karnataka	29,533	9,344	3	1228	60.5
Kerala	1,268	5,900	1	485	25.6
Madhya Pradesh	78,847	10,517	8	2152	46.9
Maharashtra	38,661	10,903	4	1938	60.1
Orissa	51,639	12,350	4	774	10.1
Punjab	12,288	6,359	2	1097	175.4
Rajasthan	35,795	5,414	7	3487	45.6
Tamil Nadu	16,600	15,844	1	438	42.6
Uttar Pradesh	124,592	22,452	6	1116	78.9
West Bengal	41,392	21,166	2	370	23.6
All India	629,364	160,529		1124	54.7

* As on 31.3.1986 Source : Fertilizer Statistics, 1986-87.

494

them) use fertilizers. This gives an indication of the need for bringing the small farmers, who are less endowed with resources, into the fold of fertilizer consumers. It was observed that the fertilizers, augmented by other inputs, support products and services, had a 'synergistic' effect and resulted in encouraging new farmers to use fertilizers. This shows that fertilizer use is more attractive, profitable and efficient when offered and used as a component of a package of improved practices. Therefore, commercial organisations should open more service centres and supply all the agri-inputs and services under one roof. As an example IFFCO started the farmers service centres (FSC) about a decade ago and now has 150 FSCs, spread over 13 states and three Union Territories of the country. Several other companies (GSFC, GNFC, NFL, etc.) also operate a network of such service centres, which have created the desired impact. The FSC supplies needed essential agri-inputs such as seed-cum-fertilizer drills, sprayers and dusters, and improved farm technology through farmers' education programmes.

Certain areas are perennially under problematic soils, while other areas are inhabited primarily by tribals. They are usually unattended to, due to some inherent problems. The development projects for these areas should be amply supported by intensive agri-input supply systems. For poor farmers, the most crucial component, out of an elaborate package of practices, should be identified. The promotion of service centres in these areas is also required.

The experience with 25 kg fertilizer packages in the hill region of U.P, the 24 Parganas in West Bengal and the tribal areas in Gujarat has been very rewarding. Small packaging will not only promote fertilizer use but will also help in increasing its efficiency since the opening of big bags for very small plots can lead to the deterioration of the fertilizer material if the farmer does not store the left-over fertilizer properly. A study must be urgently taken up (a) on the modifications, needed in the marketing and distribution of fertilizers, so that it reaches the small and marginal farmers, and in far off places, and (b) on the problem areas, alongwith other inputs, so as to safeguard fertilizer efficiency.

The prevention of losses during handling/transport, through for instance, the use of hooks, is also important. A bag has to typically suffer about fifteen handlings before it is delivered to an Indian farmer by rail.

3   Kinds of Fertilizers in Relation to Crops and Soils and New Fertilizer Materials

3.1   Nitrogen

The first series of experiments, comparing different fertilizer materials, were on urea versus ammonium sulphate, when urea was first introduced in India. These experiments were conducted for rice and wheat both at the research centres as well as on the cultivator's field. For both the crops, the two fertilizers were found to be almost equally effective.

Urea is the highest analysing (46% N) solid nitrogen fertilizer and, therefore, its transportation cost is low. It is a non-polar compound, which permits its use for foliar spray and makes it compatible to most insecticides, pesticides and herbicides. However, because of its non-polarity, it cannot be retained by clay particles and is more susceptible to leaching as compared to the ammonium fertilizer (which can be retained by the clay particles in soil). It is also hygroscopic. Surface application of urea leads to a loss of N by volatilisation. This is why some modification of urea, such as coating it with neem cake, gypsum etc. is required, to reduce N losses.

With the introduction of high yielding varieties, with their high nitrogen responsiveness, and keeping in view the low efficiency of nitrogen when applied to rice, and the high price of fertilizers, experiments were initiated at IARI and other research institutions in India to evaluate different slow release N fertilizers and the nitrification inhibitor blended ureas. These experiments showed that the efficiency of fertilizer nitrogen, when applied to rice, can be increased by about 15-25% with new fertilizer materials. Some of the promising materials are urea super granules, neem cake coated/blended urea, gypsum/rock phosphate coated urea and lac coated urea. Of these the

496

first three deserve special attention. These materials are of special relevance to the low-land rice areas of eastern India, where the top dressing of fertilizer nitrogen is rather difficult due to the standing water.

Urea super granules

The most promising of the new nitrogen fertilizers is urea super granules (USG). Its special advantage is that its production requires no additional raw material. Any urea manufacturing plant, or even a small scale entrepreneur, can install a small unit for peletting, briquetting, and granulation. IFFCO has already taken the initiative to produce USG/UB·in a pilot plant in 1978 itself. This material has been tested all over the country by different agricultural universities, research institutes and state Departments of Agriculture.

Both research experiments and field trials have been conducted and, except on some sandy soils, the results are promising. These show a 10 to 20% increase in the efficiency of nitrogen applied to rice. The Department of Agricultural Research and Education (DARE), has found USG to be more efficient than prilled urea at a number of locations (Table 11). However, despite this, the production of this material in India has not yet started. It is surprising that, in spite of experiments cver a period of 10 years, we have not yet been able to decide about the suitability or otherwise of this material, which does not require any additional raw material for its manufacture and is an efficient fertilizer. A promising material like urea super granules is therefore not yet available to the Indian farmer in the regions where it has been found to be more effective than prilled urea. Such delays are partly responsible for the low N-use efficiency on many rice fields.

Neem cake coated urea

The capability of the neem seed extract in increasing the efficiency of urea applied to rice was reported as early as 1971. Over the last 17 years, a number of experiments have been conducted at the research centres as well as on cultivators' fields. Although there is considerable variation in the results, neem cake coated urea has shown promise and

DARE has recommended this material for certain areas (Table 11) It has also been recommended by the agricultural universities in several states. However, ICAR, or the Ministry of Food and Agriculture, have taken no steps to see that this indigenous technology is effectively utilised. On the other hand, there is a good response from private agencies. A number of firms are working to develop techniques for coating urea with neem cake. IEL has made neem cake coated urea (containing 35% N) for experimental purposes. Godrej Soaps has recently come out with neem-extracted coated urea containing as much as 45% N. There are also others who would like to manufacture neem cake coated urea (one of the members of this Committee has, for

Table 11

Efficient source of N for rice and rice based crop sequences at different locations

Efficient source of N	Locations
Urea supergranules,	Bhubaneswar, Rajendranagar
Root-zone placement	Thanjavur, Varanasi, Rudrur, Navasari Banswara, Jabalpur, Kathulia, Mangalore Coastal laterites - of Karnataka, Faizabad.
Urea supergranules	Chiplima, Siruguppa, Nannaville
Root-zone placement or split doses of N with precise water control	Karamana, Palampur, Ranchi, Masodha, Raipur, Hissar.
Neemcake-coated urea	Siruguppa, Nannaville, Pantnagar,
Urea coated with Mussorie rockphosphate	Bhavanisagar

Source : Annual Report, DARE, 1986-87.

example, received a letter from a private manufacturer in Maharashtra, expressing interest in putting up a neem coating plant). However, a proper incentive, necessary registration and a mechanism to determine quality parameters, for their inclusion in the Fertilizer Control Order of the Government of India, is required to facilitate large-scale production.

Gypsum Rock phosphate coated urea

These materials were developed by the Madras Fertilizer Limited (MFL) and have shown promise during tests at research centres. Gypsum-urea products have also been developed by PDIL, Sindri and by GNFC, Baruch. Mention may also be made of lac coated urea developed by the Lac Research Institute, Ranchi.

The foregoing discussion shows that there is considerable evidence that the efficiency of urea, applied to rice (or other crops where losses are more), could be increased by modifying it, by coating with suitable materials or by briquetting or granulating.

3.2 Phosphorus

More than 50% of all P used in India is as DAP and 16% is as SSP. Moreover, over 95% of all P used in India is in area fertilizers which could most of their P in a water soluble (WSP) form.

The most important of a series of experiments, comparing phosphate fertilizers, were on nitrophosphates. Nitrophosphate (ODDA) (50% WSP) and PEC (30% WSP) were compared against superphosphate and/or DAP. These experiments were conducted under the All India Agronomic Research Scheme (now AIC.RP). Although the results varied from centre to centre, the general conclusion was that, for rice and wheat, nitrophosphates containing 50% or more WSP were required. However, for long duration crops, such as sugarcane, and for rice and on acid soils, nitrophosphates with low water solubility could be utilised equally well.

Nirtophosphates have, however, received our attention afresh primarily because they help us partly reduce our sulphur requirement. Even indigenous technology is now available for nitrophosphates. Recently PDI, Sindri, has developed the technology for making more than 90% water soluble P containing nitrophosphate rocks. Such a technology needs careful evaluation.

Phosphate rock deposits

Phosphate rock deposits are estimated to occur in the Udaipur and Jaisalmer districts of Rajasthan (nearly 100 mt) and in Mussourie in Uttar Pradesh (about 20 mt), besides smaller reserves in Jhabua in Madhya Pradesh (1.1 mt), Purulia in West Bengal (3.5 mt) and Lalitpur in Uttar Pradesh. High grade phosphate rock (about 40 mt), confined to Rajasthan, can be expected to be utilised by the industry for the production of super phosphate/phosphoric acid.

Direct application of rock phosphate to the soil is a viable proposition where (a) the phosphate rock chosen has high reactivity, (b) the soil is strongly acidic in reaction (pH less than 5.5) and (c) the crops are of a medium-long duration. Among the indigenous sources of phosphate rock, the Mussourie phosphorite can be considered to be the most appropriate for direct application. Its application could be appropriate largely in the acid soil regions of North-Eastern and South-Western India, where, incidentally, some of the plantation crops like coffee, rubber, occupy large areas.

Partial acidulation

Converting only a part of the phosphate present in the rock phosphate form, soluble in water or salt solutions, by means of mineral acid treatment, is another promising route for using indigenous phosphate rocks. The major advantage of partially acidulated phosphate rocks is in the saving of sulphur. To make this technology a viable one, the choice of the acid (nitric or hydrochloric acid alone, or in combination with sulphuric acid) and the degree of acidulation (minimum necessary) are the key determinants. The indigenous technology for this needs to be developed.

Phosphate rock-pyrite mixture

About 2 billion tonnes of iron pyrites is estimated to be present in the Amjhore deposits of Bihar. Pyrite reacts readily with air and water to form sulphuric acid, and the resultant sulphuric acid can solubilise phosphate rocks if the two materials are mixed together and a sufficient time is allowed for the reaction to take place. Though this technique holds promise, the parameters need to be carefully evaluated and standardised.

Mixture of superphosphates and rock phosphate

Utilising the ready availability of P in superphosphate and its acidic nature, a judicious combination of these phosphate sources will not only meet the immediate needs of a crop for phosphate but also cause a portion of the phosphate rock to get solubilised.

We would, however, still need to import a large amount of phosphorus. We should import the cheapest form of P per tonne, which is phosphoric acid. India has already started a joint venture with Senegal for producing phosphoric acid in that country and importing it. More of this has to be done to reduce the phosphate fertilizer costs.

We should also consider the possibility of producing higher analysis fertilizers, such as ammonium polyphosphate. A beginning has already been made by RCF and IFFCO, and we need to further it. Much will depend upon the production costs involved and a possible reduction in the transport costs. Ammonium polyphosphate also permits the inclusion of micronutrients such as zinc due to its chelating property.

3.3. Potash

There are two potassic fertilizers: potassium chloride (muriate of potash) and potassium sulphate.

Potassium sulphate, which is costlier, is only used for making cigarattes (chlorides lower the quality of tobacco used). The bulk of the potash used is as muriate of potash. All potash is imported.

The Central Salt and Marine Research Institute, Bhavnagar, developed a technology to extract potassium from the salt bitterns, as a by-product of the common salt industry in the coastal areas. The production, known as potassium schoenite, is a double salt of potassium sulphate and magnesium sulphate with about 23% K_2O and 10% MgO. An evaluation of this material, as a potassic fertilizer, in different locations has revealed that it is as good as the commonly used potassium chloride for several crops. The commercial production of this material needs to be carefully examined. The production potential is about 4% of the current total potash needs of India.

India has the world's largest deposits of muscovite mica, which contains about 10% K_2O. Mica mines produce muscovite wastes which amount to about 80% of the total quantity mined. This together with the poor quality unutilised micas, presents an inexhaustible source of potash. Scientists at the Calcutta University have recently developed a process of manufacturing potassium phosphate fertilizers by treating mica with phosphoric acid. Mixed fertilizers are produced, containing either the two macronutrients P and K or NPK. The economics of this process needs to be studied. While this process holds tremendous promise, it is essential to determine if the unreacted micaceous material has any effect on the ammonia fixation in the soil.

4   Estimates of Fertilizer Needs by 2000 AD

We had fertilizer estimates from four different sources: (a) National Commission on Agriculture (NCA), (b) Planning Commission, (c) Fertilizer Association of India and (d) Institute of Agricultural Research Statistics.

The National Commission on Agriculture (NCA) in its estimates (Table 12) has indicated a requirement of 16 million tonnes of plant nutrients by 2000 AD. This estimate is based on the food needs of the country by 2000 AD and a crop response ratio of 1:10 (nutrient: grain) for food grains. It is estimated that 75% of the fertilizer requirement will be for food grain production and 25% for non-food grain crops. NCA has also given a second estimate of 14 million tonnes of plant nutrients, if the use of HYV and other agro-techniques could increase the foodgrains response ratio from 1:10 to 1:12.5. However, the available yardstick, based on the simple fertilizer trials on the farmer's

Table 12

Consumption of fertilizers in 1970-71, 1985 and 2000 AD

	Foodgrain crops	Non-food-grain crops	Total
	(million tonnes of nutrients)		
1970 - 71	1.7	0.6	2.3
1985 trend estimate	4.5	1.5	6.0
Lower target	5.2	1.8	7.0
Higher target	6.6	2.2	8.8
2000 AD response ratio of			
1:10 for foodgrains	12.0	4.0	16.0
Response ratio of 1:12.5 for foodgrains	10.0	4.0	14.0

Source : Report of the National Commission on Agriculture, Part III Demand and Supply, page 76.

Table 13

Yardsticks of food grain production (kg grain/kg nutrient)

Crop	Levels of fertilization (Kg N+P$_2$O$_5$+K$_2$O/ha)	
	60+30+30	120+30+30
Rice	7.75	5.72
Wheat	10.56	5.78
Coarse grains	8.24	-
Pulses	7.51	-

Source : Institute of Agricultural Research Statistics, New Delhi.

Table 14

Growth in capacity, production and demand of N,P,K and K₂O - VII Plan (1985-86 to 1989-90) and 1999-2000

('000 tonnes)

Item	1985-86	1986-87	1987-88	1988-89	1989-90	1999-2000+
Nitrogen (N)						
a) Capacity	5910	6737	7021	8023	9253	
b) Production	4415	5085	5470	6231	6560	14400
c) Demand	6350	7090	7830	8570	9100-9300	
d) Gap @ (iii-ii)	1935	2005	2360	2339	2540-2740*	
Phosphate (P₂O₅)						
a) Capacity	1575	1979	2315	2453	2891	
b) Production	1319	1625	1908	2109	2190	4180
c) Demand	2200	2456	2713	2969	3000-3200	
d) Gap @ (iii-ii)	881	831	805	860	810-1010*	
Potash (K₂O)						
a) Demand	1000	1117	1233	1350	1400-1500	
b) Production	1000	1117	1233	1350	1400-1500	
Total						
a) Capacity	7485	8616	9336	10476	12144	
b) Production	5734	6710	7378	8340	8750	15580
c) Demand	9550	10663	11776	12882	13500-14000	20000
d) Gap @ (iii-ii)	3816	3953	4398	4349	4750-5250*	4420

@   To be met through imports. There is no indigenous production in K₂O.
*   Estimated (as per VII plan document).
+   Source : Seventh Five Year Plan - 1985-90. Planning Commission, Govt. of India, New Delhi.

field, conducted during 1972 to 1981, does not indicate this possibility (Table 13). The Ministry of Food and Agriculture in its Seventh Plan estimates has used a yardstick of 8 kg grain per kg plant nutrient.

The Planning Commission in its Seventh Plan Document has given an estimated demand of 20 million tonnes of plant nutrients against the estimated production of 15.58 million tonnes of plant nutrients within the country (Table 14).

The Fertilizer Association of India's estimate for fertilizer needs by 2000 AD are based on a methodology adopted by the National Informatics Centre of the Electronics Commission. They have used an exponential smoothing technique, which allows one to capture more meaningfully the impact of the immediate past. This exercise

Table 15

FAI estimates of consumption of N, P_2O_5 and K_2O (1985-86 to 1999-2000)

				('000 tonnes)
Year	N	P_2O_5	K_2O	Total
1985	6018.62	2088.08	876.78	8988.48
1986-87	6438.97	2251.58	934.94	9625.49
1987-88	6859.27	2414.75	993.10	10267.12
1988-89	7279.62	2578.93	1051.26	10909.81
1989-90	7699.88	2742.00	1109.46	11551.34
1990-91	8118.05	2905.44	1169.63	12193.12
1991-92	8538.08	3068.99	1225.81	12832.88
1992-93	8960.77	3235.91	1284.00	13480.68
1993.94	9381.05	3395.95	1341.50	14118.50
1994-95	9801.35	3559.46	1400.35	14761.16
1996-97	10641.95	3886.18	1516.69	16044.82
1998-99	11482.49	4213.35	1633.06	17328.90
1999-2000	11902.82	4376.83	1691.31	17970.96

helps us to determine the stable co-efficients (something vital given the erratic fluctuations in fertilizer demands). In this methodology, there are no independent variables and fertilizer consumption is run only on the trend variable. The time factor, in turn, automatically captures the total effect of all important variables: technology, weather, irrigation, HYVs, prices, etc. On the basis of this technique, the estimated consumption by 2000 AD is 18 million tonnes of plant nutrients (Table 15).

The Institute of Agricultural Research Statistics, in its estimate, has used a production function, based on five variables, and calculated the fertilizer needs of individual states in relation to the projected levels of food grain output by 2000 AD. These were then updated to give the country's fertilizer needs by 2000 AD. The variables taken into consideration were : unirrigated area under food grain in a state, irrigated area under food grains in a state, area under high yielding varieties in a state, the total nutrient consumption $(N+P_2O_5+K_2O)$ in a state (in thousand tonnes) and the proportion of actual to normal rainfall in the state. Assuming the the present trend continues in a linear fashion, the various input factors, including the fertilizer, for the selected regions comprising various states, has been predicted to be 13.6 million tonnes by 2000 AD (Table 16). But this would result in a production of only about 192 million tonnes of food grains, 33 million tonnes less than the projected demand of 225 million tonnes by the NCA. This means that we must raise the input levels, including fertilizers, by 10-20 per cent over the levels predicted at the current rates of growth. With a 10% increase in the level of selected inputs, the total nutrients' consumption will be 14.9 million tonnes. Similarly, assuming a 20% increase in the level of selected inputs, the estimate becomes 16.2 million tonnes. Assuming that only 75% of the fertilizer goes to the food crops, and the remaining to other crops, these values will have to be accordingly augmented. The total fertilizer consumption in India could then work out to 17 to 20 million tonnes.

All the four estimates which we examined put the fertilizer consumption demand by 2000 AD at 18 ±2 million tonnes. Considering the fact

Table 16

Projected input use and foodgrains output in 2000 AD

Regions	Gross cropped area (m.h.)	Gross irrigated area (m.h.)	Gross area under HYV (m.h.)	Total nutrient consumption (m.t.)	Food grains production (m.t.)
Alternative I					
Eastern	25.13	7.52	18.63	1.36	30.87
Western	37.10	7.22	19.17	2.09	35.02
Central	41.70	17.47	26.57	3.55	56.95
Northern	11.01	10.69	11.01	2.55	34.32
Southern	20.29	8.02	14.70	3.64	28.82
All India	*138.34*	*52.09*	*92.42*	*13.59*	*191.56*
Alternative-II (10% additional input)					
Eastern	25.94	8.34	20.49	1.50	33.70
Western	37.83	7.95	21.08	2.30	37.47
Central	43.44	19.20	29.23	3.90	62.10
Northern	12.08	11.75	11.70	2.80	37.64
Southern	21.07	8.80	16.16	4.00	30.86
All India	*143.59*	*57.33*	*101.21*	*14.88*	*207.82*
Alternative - III (20% additional input)					
Eastern	26.76	9.09	22.36	1.64	36.72
Western	38.55	8.66	23.00	2.50	39.93
Central	45.18	20.95	31.89	41.26	67.23
Southern	21.84	9.58	17.84	4.37	33.11
All India	*149.23*	*62.91*	*110.48*	*16.23*	*224.59*

Source : P.Parain, 'The Dynamics and Management of Foodgrains Production in India,' L.N. Mishra Memorial Lecture, L.N. Mishra Institute of Economic Development and Social Change, Patna.

that the Seventh Plan targets of 9.5 million tonnes for 1985-86 has not been achieved by 1987-88, for a number of reasons, one would have to take these estimates with caution. Furthermore, if sincere efforts are made, some economy in fertilizer consumption should be possible by the use of bio-fertilizers, growing legumes in cropping systems, proper accounting for residual phosphate and the use of organic manures and farm residues.

5   New Areas of Research

Fertilizer R&D

Fertilizer R&D has been a neglected area in India. This is particularly surprising since we are dependent on imported technology. Fertilizer materials, and technologies to manufacture them, have to be developed keeping in view the material resources available, the agro-ecological conditions prevailing and the cropping practices followed in the country. There is a lot to be done in this area of research; leads available must be pursued to achieve the goals.

Research on new product development should address itself to the environment identified with low fertilizer efficiency, such as wetland rice, deep water rice, coastal light soils, acid sulphate soils, vegetable farming, horticultural crops, plantation crops etc.

Products which can incorporate or retain the secondary and micro-nutrients in NPK fertilizers also merit attention.

Plant analysis, as a diagnostic tool, needs to be developed for the diagnosis of micronutrient deficiences for the orchard, plantation as well as field crops, and for suggesting measures for improvement. Good team work, involving soil scientists, agronomists and plant physiologists, would be required to reach the goal.

Cropping system Research

Research on fertilizer use on a cropping system basis has underscored the soundness of the approach, but most recommendations continue to be on an individual crop basis. Efficient fertilizer use in two cropping systems can be achieved if good soil analysis forms the basis for a fertilizer schedule for the entire cropping system, taking inter-crop adjustments into consideration. Available data show that some economy in nitrogen (20-30 kg N/ha) is possible when cereals follow a legume in the system. Green manuring with a legume may contribute 50-60 kg N/ha to the succeeding cereals. Similarly adequate phosphorus, applied to a rabi cereal, may partly meet the phosphorus needs of the succeeding kharif cereal.

The specific fertilizer recommmendations, for a cropping system as a whole, are not yet available for most situations. Fertilizer research on cropping systems needs to be strengthened, particularly in the area of phosphate management, and for systems in which crops are grown in a sequence for high yields.

The cropping system research should use mathematical modelling, to describe and predict nutrient buildings and depletions under different systems after x years at a particular level of fertilizer management. This will facilitate the transfer of results from the station to the farms. Further, because modern cropping systems include crops of different energy and market value, the performance of fertilizers in cropping systems should not only be evaluated in terms of physical-yield (tonnes/ha) but also in terms of net returns/ha and energy parameters such as Kcal/ha.

Agro-economic Research

Future needs are clearly to move towards medium/long term field experiments on well-characterised sites which actually represent substantial farming areas. Such fertilizer research should go beyond fertilizers and study the factors such as build up of pests, salinity hazard, emergence of new deficiencies etc. Traditional agronomic research should graduate into agro-economic research so that the economics of a new strategy automatically gets evaluated.

A Genetic Approach to more Efficient Use of Nutrients

For optimising fertilizer use, and increasing the fertilizer use efficiency, most research so far has been directed towards developing new fertilizer materials and manipulating agronomic practices. An alternative route, which will achieve the same objective, and will, in effect, be synergistic to the approach currently employed, is the development of plant genotypes which are inherently more efficient in absorbing the applied fertilizer, translocating it within the plant to sites of utilization, and converting it into an economic end product. Recent work has shown genotypic differences, in several crop species, in the ability to make more efficient use of the applied fertilizer. Research on improving productivity in several crop species has shown that large genotypic differences do exist in their ability to take up soil N, even when available in abundance, and also in their ability to translocate N and other nutrients to growing fruits. For example, it has been observed that the uptake of N was much higher in hybrids than in their parents. Even in legumes, which fix much of their own nitrogen, not only the soil availability of N but also its efficient utilization in the early growth phase of the plant, plays a crucial role in realising good yields. There are reports, for example, in soyabean that compensation for symbiotic deficiencies was found by more efficient exploitation of soil N, and by a more efficient redistribution of vegetative N into seeds. Further, the efficient translocation of nutrients is a major requirement for improving and sustaining productivity. Translocation is manifested through the transport of photo assimilates from the place of their synthesis, known as the source, to the growing and storage parts, known as the sink, in the plant. Photosynthesis is closely linked with biomass availability and utilisation. There are genotypic differences in biomass production at crucial stages of plant growth, and, hence, in the production and translocation of photo-assimilates. In turn, these genotypic differences can be associated with their ability to utilise the soil added N, and their ability for efficient uptake of N for biomass optimisation. There are also positive correlations reported between nitrogen uptake and seed yield. Specific differences were found in the translocation of assimilates to growing pods, between genotypes differing in their yield potential in groundnut. It should, therefore, be possible to specifically breed for

510

higher fertilizer use efficiency, and hence for various associated phenomena, for improving yields of crop plants. Thus intensive research on this important aspect will both be prudent and rewarding.

Screening of Crop Varieties

Research on screening crop varieties for nutrient stresses will be a worthwhile initiative because planting an efficient variety is half the job done. The limited research carried out so far has largely been under controlled conditions, and involving a single factor screening. Such research could be practically valuable provided (a) the green house research is backed by a screening in the field (b) the varietal screening is led from a single factor to a multifactor screening. It is known, for example, that a genotype may be tolerant to P deficiency/capable of utilising low levels of P in the medium, but susceptible to the deficiency of another element, say iron. If a really efficient variety has been identified, which can perform well under low level of nutrient application, the seed industry should be ready to multiply the seed and distribute it in the areas needed.

Mycorrhizal associations

Many plants, including crops, are known to benefit from their association with certain specific fungi (vesicular-arbuscular mycorrhiza) in the soil environment. These fungi essentially serve as an extension of the root, and forage a larger volume of soil for nutrients than possibly the root alone. By taking advantage of this naturally occurring association between certain crop species and fungi, the use efficiency of phosphate rocks can be increased to a suitable level even in soils where phosphate rock alone may not have the desired effect. This strategy is only useful on a medium-long term basis, not immediately. More research is needed.

Pollution

There is currently a global concern on the excessive use of nitrogen in crop production. In many western countries this has caused pollution of the surface and the ground water with nitrates. However,

511

the Western literature also shows that quite a bit of the nitrate comes from the disposal of cattle dung, urine etc. in intensive livestock farming. Excessive nitrates in drinking water are toxic to ruminants and human beings. Although, at the present moment, there appears to be no cause for concern in India, as our level of fertilizer and manure application is incredibly low, research avenues on this aspect should be kept open.

6   Agenda for Action

Fertilizer Research

● A mechanism to identify promising fertilizer materials, and their methods of application, carry out testing on the research centres and on the farmers' fields, and channelise for commercial production. (Action: ICAR, Agricultural Universities, Union Ministry of Agriculture, DST).

● Research and development of a simple farm machinery for application of fertilizers. (Action: CIAE/ICAR, ICAR Crop Coordinated Projects, Agricultural Universities).

● A mechanism to identify and analyse the environmental, socio-economic and operational constraints in fertilizer use and efficiency. (Action: ICAR Crop Coordinated Projects, FAI, Departments of Agricultural Economics of IARI and Agricultural Universities, STCR).

● Exploitation of the available indigenous technology in fertilizer innovation. (Action: Union Ministry of Agriculture, Department of Fertilizer).

● The creation of a centre of excellence for fertilizer research and development. (Action: DST, STR, ICAR).

Fertilizer Use Recomendations

● Integration of soil testing programmes at STCR (ICAR) and the state levels with soil survey and micronutrients coordinated projects of the ICAR, to provide a complete fertilizer recommendation service based on soil types and ecosystems. (Action: ICAR, State Governments and Agricultural Universities)

● Reorganisation of research on fertilizer use and efficiency among ICAR coordinated projects, with proper linkages with the water management coordinated programme. (Action: ICAR)

● Economically sustainable fertilizer recommendations based on the entire cropping system. (Action: STCR, ICAR, Agricultural Universities, State Departments of Agriculture).

● Provision in the ICAR Coordinated Project for long-term fertility experiments, or in the AICARP to obtain on-farm data on nutrient building and depletion under different cropping systems. (Action: ICAR).

● Research to develop plant analysis as a diagnostic tool for horticultural, plantation and annual crops. (Action: ICAR Institutions, Agricultural Universities).

● Development of fertilizer recommendations for areas with biotic constraints (pests/diseases etc.). (Action: ICAR Crop Coordinated Projects, STCR).

Fertilizer Distribution and Marketing

● Mechanism for estimating fertilizer use individually for major crops in different states. (Action: FAI, Union Ministry of Agriculture).

● Studies on the modifications required in fertilizer marketing and distribution system to promote fertilizer use. (Action: FAI, Union Ministry of Agriculture).

● Small fertilizer packages (10-25 kg) for marginal/tribal farmers and hilly areas. (Action: Fertilizer Industry, Union Ministry of Agriculture).

Policy Matters

● Mechanism for proper linkages between fertilizer industry R&D, agronomists, soil scientists and fertilizer policy makers to promote fertilizer use and efficiency. (Action: FAI,ICAR, Agriculture Universities, Union Ministry of Agriculure).

● Quicker response of the fertilizer control machinery to newly developed fertilizer products.

Infrastructure for Fertilizer Efficiency

● Provision of physical facilities for rain/irrigation water management and drainage. (Action: State Departments of Irrigation, CADA, WTC of IARI and Water Management Coordinated Projects of ICAR).

Appendix-1

Drainage of Flat Sandy Loam Soils Affected by Heavy Seepage Flow and Spring Water Under Restricted Outlet Condition
Dr S K Mukherjee

A scientifically designed drainage system, including a pumping plant, was installed in 113 ha in 1982. The project work was completed in a phased manner over a period of 3 years. A conventionally drained area covering 175 ha land under tea was taken as the control plot in these studies.

The soil of the project area has a sandy loam texture. The hydraulic conductivity is estimated to be 1.20m/day. The drainage coeffficient is estimated at 13mm/day.

The land topography is more or less flat. The slope is within 0.4 - 0.5%. The estate suffers from heavy subsurface seepage flow from large adjacent high lands under paddy cultivation and local springs. Due to the construction of a big embankment on the outlet river bank, the drainage outlet of the estate is restricted.

Based on the data on soil, crop, climate, land topography and the general hydrology of the area, a drainage system was designed. The water was discharged directly onto the artificially developed outlet drain and the flow was regulated with the help of manually operated sluice gates.

Four main drains were designed to collect drainage excess water from the project area and carry it to the outlet drain. Each main drain was provided with a sluice gate at the discharging end. A pumping plant, including a pump, a reservoir, a flat gate structure and an overhead electrification system was designed and installed. The drainage pump was of the M-F type, the inclined pump having a capacity of 1,20,000 GPH at 15' head. It consumes only 12.5 HP for operation. Piezometer wells were installed to study the water table in the project area. The water table data were recorded using an electronic water table indicator designed and fabricated locally.

The data on crop yield expert area and conventionally area drained area are presented in Table 17. Some of our observations are as follows:

Table 17
Yields as influenced by drainage of sandy loam soils having flat topography and affected by seepage and spring water

Year	Rainfall (mm)	Project area 288.36 ha					
		Experimental area (113 ha)			Conventionally drained area (175.36 ha)		
		% area under I.P.	Yield, (kg/ha) made tea	% increase in yield over 1981	% area under I.P.	Yield kg/ha made tea	% increase in yield over 1981
1985	5159	33	1953	33.8	30	1674	5.5
1984	5223	33	2015	38.0	16	1867	17.7
1983	5035	40	1826	25.1	23	1870	17.9
1982	3832	Nil	1824	24.9	28	1859	17.2
1981*	4411	9	1460	-	31	1586	-
1980	4450	61	1709	-	19	1795	-
1979	5075	26	1808	-	33	1893	-

*Pre-drainage year.

● From 1978 to 1981, the estate faced a steep declining trend in crop yield.

● The old existing conventional drainage system in the control area was maintained as it was. This included a lateral drain two inch deep and with a 20-30' part in the east-west and the north-south directions.

● The increase in yield recorded from the experimental area was 24.9, 25.1, 38.0 and 33.9% in 1982, 1983, 1984 and 1985 respectively over that of the 1981 yields (pre-drainage year). This increase in yield was obtained in spite of the fact that the area under LP was 40, 33 and 33% in 1983, 1984 and 1985 respectively in the experimental plot. This means that 106% of the experimental area, as against 769% in control area, was given LP during the last 3-year period. Due to certain problems, the drainage pumping plant could not be operated during the 1985 rainy season. This caused a small little setback in the crop yield in 1985. A fine plucking standard maintained may also have caused some depression in the total yield during 1985.

● The crop yield data from the control plot show that, after an initial increase of 17.2% in 1982, the yield remained same in 1982, 83, and 84, i.e. there was only 8 kg/ha increase in yield in 1984 as compared to the 1982 yield. In 1985, there was a sudden decline of 1.93 g/ha as compared to 1984. On an average, there has only been a 88 kg/ha increase during the last four-year period as compared to the base year crop i.e. 1981. In the same period, the experimental area recorded an increase of 4.93 g/ha, as against 88 kg/ha from the control plot, which is quite remarkable.

Appendix-2

Rice Cultivation in Eastern India

Eastern India, constituting the states of Assam, Bihar, Orissa, West Bengal, Eastern MP, Eastern UP and the North-Eastern states of Manipur, Tripura, Nagaland and Arunachal Pradesh have 18.24 m. ha under rice. about 46.8% of the total rice area of the country. The average paddy yields in this region are 0.5 to 2 t/ha, depending upon the ecosystem. This is because the crop is mostly grown under rainfed conditions with virtually no control over the amount and timing of water supply. The traditional rainfed rice culture has different kinds of areas: (a) uplands, (b) shallow lowlands (5-15 cm standing water) and (c) deep water rice (above 100 cm standing water).

Upland rice is grown in areas with an annual rainfall of above 800 mm during the crop season. The land, whether slopy or flat, bunded or unbunded, may have no standing water on the soil surface 48 hours after cessation of rain. The problems encountered in upland rice are: drought, surface run-off, deep percolation, weed infestation, high soil acidity, low availability of nutrients like P, K and Ca and, at times, Al and Fe toxicity.

Lowlands (shallow, intemediate and deep) are converted into flooded paddies with the onset and the intensification of rainfall in July. Flood plains of the vast river systems in the eastern and north eastern India are mostly lowlands. There is usually high rainfall intensity during late July to mid-August, and the water accumulates to different heights because of the rise in the ground water table, flash floods and the slow drainage of the surface water. Top dressing with fertilizer is not possible in the rainfed lowlands. These areas would need fertilizers that can stay in the soil for a long time so that, when applied at sowing/transplanting, nutrients are available throughout the growth period of rice.

In situations where small or sub-optimal amounts of fertilizers are used, the strategy should be to increase their application rate to a total of 20-40 kg N/ha and P and K based on soil test data. In areas where sufficient or reliable soil test data are not available, the strategy should be to add at least 20 kg/ha each of P_2O_5 and K_2O in addition to the form, method or time of application.

Split application of N is the cheapest and most widely used practice followed by farmers. Application of 100 kg N/ha in splits can raise grain yields of 3000-1200 kg/ha compared to the same quantity applied basal in a single dose at one time. This technology is feasible for most of the dryland areas and irrigated situations with good water control. In irrigated and rainfed areas, particularly under rice cultivation, where no control of water can be exercised, pre-treatment of urea with moist soil can make split application more remunerative.

Appendix-3

Availability of Organic Manures in India

The zonewise rice area of India under different land situations and the potential of organic and biological resources and plant nutrients are shown, respectively, in Tables 18 and 19.

Table 18

Zonewise rice area (million hectares) of India under different land situations

Zone	Irrigated	Rainfed upland	Rainfed lowland including deepwater rice	Total
East zone (Assam, Bihar, M.P., Orissa, Uttar Pradesh, West Bengal)	7.59	5.23	14.61	27.43
North East hill states (Arunachal Pradesh, Manipur, Meghalaya, Nagaland, Sikkim, Tripura)	0.23	0.28	0.27	0.78
North zone (Jammu & Kashmir, Punjab Haryana, Himachal Pradesh)	2.51	0.07	-	2.58
West zone (Rajasthan, Maharashtra, Gujarat)	0.71	0.79	0.75	2.25
South zone (Andhra Pradesh, Karnataka, Kerala, Tamil Nadu)	6.58	0.64	0.73	7.95
Other small areas	-	0.08	0.09	0.17
All India Total	17.62	7.09	16.45	41.16

Table 19
Potential of organic and biological resources and plant nutrients

Name of resources	Annual potential (M.T.)			Plant nutrients (M.T.)			Total (M.T.)
	Dung	Urine	Biomas	N	P_2O_5	K_2O	
Cattle	744.565	480.148	-	2.977	0.793	1.332	5.102
Buffalo	258.022	178.753	-	0.745	0.276	0.487	1.508
Goat and sheep	12.228	7.918	-	0.214	0.063	0.020	0.297
Pigs	4.596	3.990	-	0.044	0.027	0.029	0.100
Poultry	3.395	-	-	0.027	0.020	0.010	0.057
Other live-stock	6.024	4.095	-	0.079	0.018	0.069	0.166
Human beings	30.380	274.100	-	3.228	0.776	0.715	4.719
Crop residues	-	-	100	0.500	0.600	1.500	2.600
Forest litter	-	-	15	0.075	0.030	0.075	0.180
Water hyacinth compost	-	-	3	0.060	0.033	0.075	0.168
Rural compost	-	-	226	1.130*	0.678*	1.130*	2.938*
Urban compost	-	-	6	0.024	0.015	0.030	0.069
Sewage sludge	-	-	0.3	0.012	0.009	0.003	0.024
Rhizobium	-	24m	@30-400 Kg N/ha	1.00	-	-	1.00
Non-legumes	-	10m	@35Kg N/ha	0.15	-	-	0.15
Blue-green algae	-	30m	@25Kg N/ha	0.75	-	-	0.75
Total				9.885	2.660	4.345	16.89

*Total excludes contribution of rural composts since it is prepared from animal excreta and crop wastes. These resources have been included.
Source: Gaur, A.C. Neelakantan, S. and Dargan, K.S., (1985) I.C.A.R. Tech. Bull on Organic Manures.

11
Future Food Needs

In the ultimate analysis, the packaging of the best available technologies into an economically and ecologically desirable farming system should be the end point of all scientific effort. To achieve this, we must promote the optimum use of all the available resources: land, water, climate, labour and credit.

Contents

11
Future Food Needs

Preamble

In India the per capita availability of land for agriculture at the end of the nineties will be about 0.15 ha. The available land will have to provide the food, fibre and other needs of about 100 crore men, women and children, and fodder, feed and other needs of about 50 crore farm animals. This task has to be accomplished in an agricultural environment characterised by a steady loss of the biological potential of the soil, and by an increasing impoverishment of the biological wealth of the country. The dimensions of the scientific challenges ahead, for concurrently achieving ecological and food security, are probably unprecedented in human history. The scenario has to be viewed in the light of our past achievements so that effective and rational ways can be devised for facing the challenge. ∎

1   Introduction

1.1   The Achievements

The annual growth rate in food grain production in India between 1900 and 1947 was hardly 0.1 per cent. After Independence, we can recognise two broad phases in our agricultural evolution. The first phase, starting from the beginning of the First Five Year Plan in 1950-51 and ending in 1966-67, was marked by considerable progress in the development of the infrastructure, essential to offer farmers mutually supportive packages of technology, services and public policies.

Agricultural production increased during this period largely through an expansion in cultivated and irrigated area.

The second phase, starting with the wheat harvest of 1967-68 and ending in 1988-89, has been characterised by productivity improvement as well as a relatively greater stability in total agricultural output, particularly during the season. The introduction of high yielding genetic strains of wheat, rice, poultry, dairy cattle, fish and other plants and animals, helped to maximise the benefits from better nutrition of crops and animals and from water. The chemical industry played a key role in initiating and maintaining the momentum of progress. During 1951-86, the consumption of nitrogenous fertilizers increased nearly 100 times, from 58000 tonnes to 5.81 million tonnes. The most important contribution of the second phase was to create the confidence that India can, with the help of its farmers and scientists, build a stable national food security system based on home-grown food. The nineties will mark the beginning of a new phase where our agriculture will face serious challenges on the ecological, economic, energy, equity and employment fronts.

Scientists will have to struggle with problems arising from the loss of the biological potential of the soil, on the one hand, and the biological impoverishment of the country, on the other. Economists and policy makers will have to struggle with the problems arising from the increasing diminution of land holdings and from the cost-risk-return structure of small farm agriculture. Social scientists will have to find means of ensuring that the benefits of new technologies reach the unreached. Technologists will have to work out optimum energy mixes for agriculture in different parts of the country. They have to help also in generating more opportunities for skilled employment in villages through symbiotic linkages between the farm and off-farm sectors. Post-harvest technology has to receive much greater attention. Political leaders have to enlarge the priority for rural infrastructure development and develop a new vision for our agriculture as a source of income and employment, and not just as a source of food for the growing population. They have to give priority to improving the purchasing power of the rural and urban poor, *since economic access to food is already the most important food security challenge facing the country*. Above all, they must work hard to promote a population policy which will help to stabilise our population at a level that the carrying capacity of the available land and water resources can support on an ecologically sustainable basis.

1.2   Science and Technology in Agriculture

To promote food production on a sustainable basis, we need mutually supportive packages of technology, services and public policies.

1.2a   Package of Technology

Farming systems and practices have evolved over centuries, and what are usually referred to as traditional farming systems have evolved through observation and experience to suit specific agroecological conditions. Such systems, however, are generally not intended to raise crops or farm animals for the market. When the process of modernisation sets in, the principal catalyst is the opportunity for remunerative marketing. When food crops are grown for the market, the distinction between them and traditional cash crops disappears. For example, in several parts of India, wheat or rice is the main cash crop for numerous farming families.

Market-oriented agriculture succeeds only with a high degree of efficiency both at the production and postharvest stages. The goal is to achieve the highest yield possible per unit of land, water, time and labour. This is where improved technologies become essential.

During the first 20 years after India's independence, production advances were achieved largely by increasing the area under cultivation. With the pressure of a growing population, this pathway of increasing production could only lead to a dead end. Improved productivity, and intensitified cropping, became necessary to produce the food needed for a growing population.

One of the first tasks, therefore, was to build a scientific infrastructure to stimulate and sustain rapid agricultural advance. Agriculture is, by and large, a location-specific vocation; hence, a dynamic national research system is a must for sustaining a dynamic production program.

India is fortunate to have an excellent network of agricultural research institutes and universities. At the national level, the Indian Council of Agricultural Research (ICAR) supports and coordinates scientific research, training and extension education in crop husbandry, horticulture, animal husbandry, fisheries and agroforestry. ICAR is a unique organisation in that it has concurrent responsibility for both research and education. Unfortunately, in most developing countries, the responsibilities for research, extension and education are divided among different agencies and ministries.

An important element in developing strong research organisations is the personnel policy which must be designed to attract and retain dedicated and high-quality staff. ICAR introduced an agricultural research service (ARS) in 1974 to encourage specialisation and to provide job security and opportunities for promotion, even without the occurrence of vacancies. A National Academy of Agricultural Research Management was set up in Hyderabad to provide opportunities for in-service training to the members of ARS as well as to promote a scientific culture within the organisation.

Research programmes in India are carried out in a large number of central and state research institutes and national bureaus such as the National Bureau of Soil Survey and Land Use Planning and the National Bureau of Plant Genetic Resources. Linking the agricultural universities and central institutes are the All India Coordinated Research Projects. These projects bring together scientists working in different institutions and disciplines in a symbiotic partnership. The individual strengths of the different research institutes may vary, but the collective strengths of All India Coordinated Research Projects are considerable.

Once improved technologies are developed by scientists, the experimetal findings are verified in the farmers' fields before they are recommended for widespread adoption. ICAR has developed the following methods of demonstrating and verifying new experimental findings:

- The "lab to land" programme, which aims at extending new experimental findings to small farmers

- National demonstrations, which are designed to show that small farmers can achieve high yields, and

- Whole-village or watershed operational research projects, which can help identify the major constraints responsible for the gap between potential and actual farm yield and income.

Since illiteracy is still widespread in India, ICAR has organised a network of Farm Science Centres (Krishi Vigyan Kendras) all over the country for disseminating the latest technical skills through a learning-by-doing scheme.

Agricultural research institutions and universities have identified themselves with the farmers, who often spend many hours visiting the experimental fields and discussing, with scientists, the problems of mutual interest.

Technology development has been used not only on individual factors of production but also on farming system as a whole. In summary, ICAR has been able to achieve the following:

- A close integration of research, training and extension education

- A national grid of cooperative experiments bringing together scientists from different disciplines in a working partnership to achieve specific research goals

- Close linkages between research and development agencies, and

- Farmers as partners in the testing and refinement of technologies.

How the research system responded to a specific challenge is illustrated by the following examples. After Independence, fertilizer factories were established and production began, both in the private and public sectors. Fertilizers use in farmers' wheat fields was started in the 1950s, using low doses of about 20 kilograms of nitrogen per hectare. Studies showed that there was no economic response to this low level of nitrogen application. Hence, in the early 1960s, research on the breeding of varieties that could respond well to water and fertilizer application was begun. The semi-dwarf wheat varieties from Mexico, introduced through the help of Dr N E Borlaug, provided suitable strains that could repond well to a good soil fertility management.

A National Demonstration Programme was started in 1964-65 to illustrate new opportunities for improving productivity by using management-responsive varieties. A small government programme was soon converted into a mass movement by farmers, and, as a result, the area planted to semi-dwarf varieties increased from about 4 hectares in 1964 to about 4 million hectares in 1971. From a peak production of about 12 million tonnes in 1964-65, before the introduction of high yielding varieties, wheat production in 1985-86 exceeded 46 million tonnes (Tandon and Seth 1986, p17 and Government of India, 1985, p2).

India's progress in wheat production is well-known. However, it is important to realise that such rapid progress was possible only because of the existence of a dynamic national research system with the capacity to derive maximum benefit from international research centres like the International Centre for Maize and Wheat Improvement (CIMMYT) in Mexico.

The second example is of the recent development in rice production, which accounts for about 40 per cent of the total food grain production in India. Like grain production, rice production was stagnant until the mid-1960s, when new genetic strains were introduced that were able to respond to good soil fertility and water management. Rice is predominantly a crop of the southwest monsoon period (May to

October) when problems of pest and water control make it difficult to increase yield. Fortunately, the All-India Coordinated Rice Research Project of ICAR, which works in partnership with IRRI, helped to provide location-specific varieties and technologies for different parts of the country. Because of new technologies, non-traditional rice areas, like the Punjab, and non-traditional seasons became important in rice production. The government's provision of adequate energy to pump water from tubewells has enabled Punjab to maintain high rice yields even in severe drought.

To sustain yields at high and stable levels, effective monitoring and early warning systems are needed, particularly with regard to soil conditions and plant and animal health. In agriculture, conditions are dynamic, and constant vigilance is necessary. To cite an example, there was virtually no economic response to phosphorus application in Punjab in the 1960s. In the early 1970s, however, phosphorus application became essential to obtain substantial returns from applied nitrogen. In a few years, zinc deficiency became widespread. Thus, when farming practices shift from low to high productivity levels, new problems of plant and soil health may arise, making a location-specific research capability vital.

The national goal of agricultural research is the improvement of the productivity, profitability, stability and sustainability of the country's major farming systems. Sustainability has to be viewed from the ecological as well as the economic angle. We cannot allow depreciation of basic agricultural assets like land, water, flora and fauna. While ecological sustainability is fundamental to the future of agriculture, economic sustainability is vital to generate the interest of farmers in increased production. An effective national agricultural research system should look at both ends of the spectrum. Improving productivity without detriment to the long-term production potential of the soil, and creating economic viability that results in a satisfactory "take-home income" for farmers, are both essential. To succeed in both these areas biologists and social scientists must work together as a team.

1.2b   Package of Services

Applied research is no different from 'ivory tower' research if the resulting new technologies are not transferred to the farmers' fields. In most countries of South and Southeast Asia, where the average size of a farm holding is usually less than one hectare, the cooperation and assistance of at least a million farming families is needed to produce an extra million tonnes of grain. No production target can be achieved without first asking the question "why should farmers produce more, and how will they do it?"

If this question is asked at the outset of a development project, priorities can be established to ensure that services, that enable farmers to take advantage of new technologies, are provided. In India, both government and private agencies have been active in providing inputs iike seeds, fertilizers, pesticides and, very importantly, credit. Government policies in the development of effective input supply services have evolved over a period of time. Public sector companies like the National Seeds Corporation and credit institutions like the National Bank for Agricultural and Rural Development have become very important instruments in India for enabling small farmers to adopt new technologies.

The mass media, particularly radio and local language newspapers, have been extremely important in the disseminauion of agricultural production. Television increasingly is becoming a powerful communication tool. Extension services, including the Training and Visit (T&V) system of structured knowledge transfer, have grown in capacity and effectiveness.

The major goal of the input supply system is to render new technology accessible to all farmers. Ensuring equality of access to appropriate technology should be the foundation of agricultural extension and development planning. Knowledge and skill transfer systems, and methods of providing inputs needed to convert knowledge into field accomplishment, must be synchronised in space and time. Input delivery systems must be tailored to specific sociocultural conditions.

In India, several innovative systems of input delivery to illiterate farmers have been developed by both governmental and non-governmental agencies. For example, credit fairs have been organised to eliminate red tape and corruption, and to ensure that credit is utilised for the intended purpose. The goals of a credit fair are the following:

- Identify a well-designed production or input supply programme such as the construction of tubewells, farm ponds or other minor irrigation elements, or the supply of fertilizers, seeds and other production inputs.

- Specify the criteria for credit eligibility, including the amount of credit and conditions under which it will be made available.

- Organise a tie-in between credit supply and the supply of inputs for which the credit is intended.

- Announce, three to four weeks before the date of the fair, the details of the programme, including the location and conditions for eligibility.

- Organise teams of fair managers consisting of representatives from the banks and input supply agencies. They should have a short training programme to prepare them to supervise the smooth operation of credit and input supply. In addition to the regular extension personnel, volunteers from agricultural universities and institutions could be mobilised to ensure that the program is carried out efficiently and effectively.

In 1975, the National Commission on Agriculture issued a comprehensive report reviewing the progress in Indian agriculture since 1927, and suggesting methods of meeting future challenges. It recommended the organisation of country-wide farmers' service societies to provide farmers with credit and the inputs for which the credit is intended, all under the same roof (Government of India, 1976, pp. 256-83).

In recent years, the availability of key inputs in remote areas has been monitored rigorously. A checklist of all essential activities is verified at the village level during a specified two-week period, usually several weeks before the date of sowing. This system provides an early warning of potential critical bottlenecks so that they can be removed before the onset of the sowing season.

1.2c Package of Government Policies

Even if good technologies are available and the input supply system is efficient, farmers will be able to derive benefit from them only if some basic steps are taken by the Government. The first area of action is in the realm of land reform. Farmers need to have a long-term stake in the land to make them invest in the infrastructure essential for sustained productivity. Security of tenure, land ownership patterns and the size of the farmholdings need particular attention.

In the past, farmers based land-use decisions largely on the needs of both the family and the immediate neighbourhood. With the modernisation of agriculture, farmers produce food grains and other commodities not only for themselves, but, more importantly, for the market. When this transition takes place, opportunities for producer-oriented and remunerative marketing become an important factor in sustaining and stimulating farmers' interest in modern technology.

Input-out pricing policies become crucial. The turning point in Indian agriculture took place in 1964, when the Government set up a commission to recommend minimum prices for food grains, in order to provide an incentive to farmers, and a good corporation to assure that farmers received those prices. The Government has honoured its commitment to farmers by purchasing all surplus grain offered to government agencies. Because of this, the Government now has fairly substantial grain reserves.

Symphonic and Knowledge-Intensive Production Systems

The term "symphonic agriculture" designates the evolutionary stage reached in the development of sustainable agricultural production systems, when all of the components of an agricultural action plan become mutually reinforcing. When synergetic packages of technology, services and public policies are developed and introduced in a mutually supportive manner, agricultural progress is rapid. Further, government investment policies ultimately determine the fate of rural professions. Without adequate rural infrastructure such as roads, warehouses, electricity and facilities for education and health care, rural areas will not attract technically qualified people. In fact, the migration of educated and well-trained people from villages to towns is the most serious form of brain drain in most developing countries. In India, as in most developing countries, youth below the age of 21 constitute the majority of the population. Thus, the future of agriculture, and of rural professions, will depend on how we make agriculture both economically and intellectually attractive. This calls for greater effort in developing and popularising knowledge-intensive production systems. At the same time, the minimum essential living facilities, and the opportunity to acquire consumer goods, must be provided in rural areas.

Essentially the message from our past experience is the following:

- Develop a strong national research capability in agriculture that can help to provide optimum returns from the land, water, livestock and capital resources.

- Develop methods of transferring knowhow and skills to farmers and provide them with inputs necessary for converting technological advances into production gains.

- Introduce agrarian reform, rural development and communication concepts, input-output pricing policies and other programmes that can stimulate and sustain the growth of market-oriented farming.

- Stimulate consumption by the rural and urban poor through various measures including food-for-work and employment guarantee projects.

535

- Consider agriculture not just as a means of producing food for the urban population but also as a powerful instrument for increasing income and employment. Such an approach is essential to enhance the purchasing power of the rural population. This calls for a labour-friendly approach in technology development and transfer.

These components of an agricultural action plan are the basis for a sustainable agricultural production system.

2 The Challenges

2.1 Nature of Challenges

Social Factors

The average size of a farm holding is decreasing. Over 9 crores out of the present 10 crores operational holdings, belong to the small and marginal farmer categories. The number of landless labour families is increasing. Women usually perform jobs involving much drudgery and little remuneration. Youth constitute the majority of the rural population and the educated among them are not satisfied either with the infrastructure available in villages for health care and education, or with the prevailing opportunities for skilled employment. Consequently there is a drain of brains from the village to the city and the town.

Economic Factors

Small farm families are affected more by the cost-risk-return structure of farming than the more well-to-do sections of the farming community. For them stability of income is as important as yield per hectare. There is a mismatch betwen production and post-harvest technologies, particularly with regard to perishable commodities. Knowledge-intensive techniques, which can reduce the cost of production without reducing yield, become particularly relevant. Also, income from farming alone will not be adequate to meet the needs of the family for a better quality of life. Farm income will have to be supplemented by off-farm income.

Energy

Without adequate availability of energy sources, productivity and efficiency cannot be improved. There is need for more power for the farm sector, if the efficiency of the farm sector and farm operations, at the production and post-production stages, is to undergo any marked improvement. Energy is particularly needed in dry and drought-prone areas for the timely preparation of land for sowing and for taking advantage of ground water resources. An integrated energy supply system, involving both renewable and non-renewable sources of energy, needs to be developed for every block.

Employment

The farm sector plays a dominant role in providing employment. In order to attract and retain educated youth in agricultural occupations, farm technologies should be both intellectually satisfying and economically rewarding. A knowledge-intensive agricultural system, which can also generate considerable downstream employment in the secondary and tertiary sectors, will be necessary to create more skilled jobs for rural men and women. An integrated approach to on-farm and off-farm employment, and to blending traditional and frontier technologies, is urgently needed. For this purpose, the on-going farming systems research (FSR) programme will have to be modified and converted into a rural systems research (RSR) programme, specif. cally designed to generate better growth linkages (Swaminathan 1988).

Ecology

In the ultimate analysis, only ecological security can safeguard both food and water security and livelihood security. The conservation and sustainable management of land, water, flora, fauna, and the atmosphere are essential for safeguarding the future of our agriculture. The unique coastal, mountain, wetland and dryland ecosystems of our country need urgent attention from the point of view of conservation and rehabilitation. Sustainable land use systems will require detailed attention with reference to the following three categories of land:

- land which needs to be conserved in its pristine purity because of its rich endowments of forests, wild life and other forms of biological diversity

- land which is degraded and which needs to be upgraded through appropriate techniques of restoration ecology. These lands are usually referred to as "waste lands" and,

- land which can be subjected to sustainable intensification through scientific agriculture and aquaculture.

Social Engineering

In view of the small size of holdings, and the limited financial and technical resources available to small farmers and agricultural labour families, it is essential that the organisational aspects of small farm agriculture receive adequate attention. A small farm has great potential for intensive agriculture, but a small farmer is usually confronted with serious economic problems arising from the cost and risks involved in intensive farming. *It is only by helping the small farmer to overcome his/her economic and technological problems that the potential of a small farm for high productivity can be realised.* Also, knowledge-intensive and ecologically sound agricultural practices, such as scientific water management, integrated pest management, and improved post-harvest technology, require, for their efficient adoption, cooperation among a group of farmers living in a village or watershed or the command area of an irrigation project.

We can realise the full potential of small farm agriculture if individual initiative, group cooperation and government support can be promoted in a mutually reinforcing manner. Group insurance schemes, group subsidy, group oriented extension and other services and incentives for group-cooperation need to be devised and introduced quickly.

538

We should develop organisational structures which can promote decentralised production supported by a few key centralised services. Special attention is needed in the case of perishable commodities for promoting producer-oriented marketing. For example, it will be useful to promote the organisation of horticulture estates, poultry estates and aqua-culture estates, to provide the needed infrastructure facilities and services to small producers of vegetables, fruits, flower, eggs, broilers and fish. The features already found useful in the dairy sector need to be transferred to other sectors. *Social engineering techniques should ensure that new technologies reach the unreached and, at the same time, help to promote the growth of a socially relevant and dynamic services sector which can provide skilled jobs in villages to the young men and women belonging to families without productive assets.*

2.2 Meeting the Challenges

The following are some of the principles which should form the basic guidelines for the agricultural strategy for the nineties:

● *Land, water, energy and cost saving crop husbandry,* i.e., productivity advance and water and energy use efficiency become the major pathways for enhanced production

● *Grain saving animal husbandry* i.e., the avoidance of animal husbandry practices which need large quantities of cereal grains to feed animals and the initiation of a programme for the establishment of block level biomass refineries

● *To the factory but not to the town* i.e. the generation of adequate number of skilled jobs in the secondary agro-processing and rural industries sector, and the tertiary services sector, so that those without assets or work need not migrate to towns and cities

● *Sustainable management of the basic agricultural assests* of land, water, flora, fauna and the atmosphere: for this purpose, current patterns of productivity measurement should be changed, as will be pointed out later

- *Income and employment orientation* to agricultural research and development so that agriculture becomes not only an instrument for the production of more food grains, and other commodities, but also a catalyst for the production of more jobs and income in rural areas.

3 Land

It would be useful to divide the country into the following science and technology land zones:

Green revolution areas

These areas occur all over the country, and are characterised by substantial improvements in the productivity of crops, particularly wheat and rice. A major challenge in these areas is to defend the progress already made and raise the ceiling to yield further. This will call for the introduction of frontier technologies including the development and cultivation of hybrid rice. Post-harvest technology will also need additional attention so that all links in the producer-consumer chain are attended to. Through cereal-legume crop rotations, animal husbandry programmes can be supported at a higher level of efficiency of stall fed animal rearing.

Green, but no green revolution areas

Most of eastern Indian comprising north Bihar, eastern Madhya Pradesh, north Orissa, parts of eastern Uttar Pradesh, West Bengal, Assam and the north-east States fall under this category. Here, availability of water is not the primary constraint and the land is green most of the year. Underground water is yet to be tapped extensively. Scientific management of water and nutrients holds the key to improving productivity in these areas. New methods of feeding the crops during the southwest monsoon period are needed, since the loss of applied nitrogen in the form of urea is generally over 60 per cent. In the coming decades we can maintain a stable food security system in the country only if we are able to enhance productivity in this region on a sustainable basis.

Dry land farming

Saving and sharing water, new approaches to breeding and feeding of crops, agro-forestry systems, with particular reference to leguminous shrubs and trees, animal husbandry and off-farm employment: all these aspects will have to receive much greater attention. Steps to tap ground water on a sustainable basis, and the development of alternative cropping strategies based on different rainfall and moisture availability patterns, are needed.

Arid Ecosystems

Both the hot desert of western Rajasthan and the cold desert of Ladakh provide unique opportunities for high quality seed production, arid horticulture and animal husbandry. Ladakh could be the venue for a national genetic resources repository, operated at a low cost under permafrost conditions.

Mountain Ecosystems

The Himalayas, the western and the eastern ghats and the Vindhyas provide great opportunities for the production of high value crops, including plantation crops, medicinal plants, livestock products and catchment area development. Deforestation in mountain eco-systems is causing great harm to downstream agriculture, in addition to causing the loss of top soil and biological diversity.

Coastal area agriculture

The coastal areas afford opportunities for integrated agriculture and aquaculture production systems. Also coastal agriculture can be given an export orientation since transport costs can be minimised.

Island Ecosystems

The Andamans, Nicobar and Lakshadweep groups of islands afford unique opportunities for nurturing biological diversity, establishing

off-shape quarantine facilities and high-tech aquaculture, in addition
to eco-tourism. Marine national parks can be established at suitable
locations.

Wetlands

Detailed planning is necessary for the sustainable management of wet
lands to produce a variety of adequate products and to avoid biological
impoverishment.

Waste lands

The rehabilitation of degraded lands has to be undertaken using
technologies most appropriate to each situation. An energy effort
should be made to restore to wastelands their full biological potential
by the year 2000.

4   Water

Our population will be over a billion in the early part of the 21st
century. The conservation and scientific use of water must occupy a
very high priority in the national agenda for sustainable development.
The areas which need urgent attention are:

Irrigation Extension

There is a vast pool of know-how already available on water management. Yet, the application of this knowledge, at the field level, is poor.
A *group endeavour* , on the part of the water-shed or command-area
or catchment-area community, is essential for efficient water use. We
need an efficient extension system which can promote both individual
and group action for the economic and effective use of the available
water.

Area Specific Water Management

Waterlogging is a major problem in the irrigated areas of north Bihar
and several parts of eastern India during the *kharif* season. Drainage

and ground water use during *rabi* could improve both the yield as we l as the infiltration capacity of the soil in Punjab, Haryana and north-west Uttar Pradesh. On-farm water management and energy management will hold the key to improved production in these areas. The precise priorities in water management will vary from region to region.

Farming System Specific Water Management

A specific water management system will have to be tailored to each farming system. In the design or farming systems, the optimum use of the available water should be a major aim. Farming systems may involve multiple cropping, crop-livestock integration or a crop-fish combined production system. The productivity of a farming system can be improved if the soil-water-plant relationships are studied in an integrated manner.

Aquifer Management

It is high time we develop, with the cooperation of farmers, a scientific ground water management strategy. This should ensure that we do not live on our capital. In chronically drought-prone areas, "ground water sanctuaries", on the lines proposed during the drought of 1979, deserve special attention.

Problem Soils

Water management in alkaline and saline soils and heavy clay soils needs greater attention in relation to the farming system in use.

Management of Irrigation Commands

Unless the management of command areas of irrigation projects and water sheds becomes a joint sector activity, based on shared goals and perceptions between farmers and Government officers, the existing inefficient, and often scientifically harmful, methods of water use will continue.

Operational Research Projects on Water Management

Operational research projects involving irrigation departments, agricultural universities and institutes and farmers, should be designed to demonstrate the value of *Pai Panchayats* (Water Cooperatives), the impact of improved water management on farm income and the need for adequate maintenance mechanisms.

Generating Consciousness of Water Use Efficiency

In many parts of our country, the cereal productivity per unit of water can exceed 1 kg/m^3 with efficient water management. However, in practice, it is usually less than 0.4 kg/m^3. Agronomists should express yield not only in terms of kg/hectare but also in terms of kg/cubic meter of water. There is also a need to sensitise farmers and administrators on the potential adverse effects of improper water use. Salinisation, waterlogging and the spread of waterborne diseases such as malaria and filariasis can be avoided if anticipatory steps are taken.

5   The Ecological Dimension and Food Security

Our chemical industry spearheads the biotechnology movement and links it with ecological and food security. Ecological security implies the conservation and sustainable mangement of the basic life support systems of land, water, flora, fauna, and the atmosphere. It involves concurrent and integrated attention to all the components of the biosphere and geosphere. Ecological security is the only foundation on which enduring edifices of food and livelihood security can be built. The ecological security of our planet is under threat today from human life styles, the patterns of agricultural, industrial and economic developments and urbanisation. Population growth is exceeding the capacity of natural ecosystems to support them on an ecologically sustainable basis in several parts of our country. Increased emissions of carbondioxide and other "greenhouse" gases arising from various human activities are leading to the phenomenon referred to as global warming. Deforestation is leading to a loss of biological diversity. These threats to the quality of life on earth are leading to a more widespread awareness of the fact that nature's patience is not inexhaustible.

A disregard for the long term consequences of current patterns of development and population growth are leading to:

- air and water pollution
- soil erosion and land degradation
- deforestation and changes in microclimate
- loss of biological diversity leading to enhanced vulnerability to pests, pathogens and abiotic stresses such as salinity and alkalinity
- migration of the rural poor to towns and cities as a result of the loss of their livelihoods in their native villages (some times referred to as "environmental refugees").
- global warming, acid rain and increased ultraviolet radiation.

To reverse these trends, the World Commission on Environment and Development, chaired by Mrs Gro Harlem Brundtland, Prime Minister of Norway, has, in its report titled *Our Common Future*, urged all nations to accord the highest priority to making development ecologically sustainable.

While it is important to think about the future, what is even more important in our country is to plan for *a better common present*. The disparities in the lifestyles of the poor and the well to do, and in the amenities available in rural and urban areas, are ever widening. It is in this context that we should examine the opportunities opened up by new technologies.

6 Biotechnology

Biotechnology has come to be regarded as the supplemental innovation that the conventional plant breeding methods so very badly need for developing improved crops. The attractions are: (a) a more directed means of creating gene pools enriched with desired traits, (b) a more efficient and predictable set of selection criteria and (c) the ability of assimilating desired genes from pools not accessible through conventional methods. In recent years, molecular biology has provided unprecedented insights into basic mechanisms of biological processes, and has opened up possibilities of unlimited flow of genes and the power of regulating genetic expression. Expectations from new biology in the industrialised world are high. Naturally, the developing countries do not wish to be left out of the bonanza.

6.1   Options

For transition from the conventional to the modern methods for crop improvement, two major options are considered for the developing countries (a) strengthening of the conventional programmes undiluted by distractions of hi-tech science. It is argued that the developing countries are a very heterogenous group with regard to the strengths of the on-going crop breeding activities. Since such an activity is crucial in its own right, and is an essential pre-requisite for injection of modern crop improvement methodologies, the developing countries need not make any investment; they should instead wait till the fruits are ready to pick and then borrow the technology and exploit it. (b) The other option is to develop a research base in molecular biology and genetic engineering simultaneously with the development of conventional breeding.

The choice of the optimal operational option will have to relate to the state of scientific development in the country and the quantum of resources that can be set apart. In countries where even conventional plant breeding programmes do not exist, research on molecular biology and genetic engineering for crop improvement will be unprudent. However, in developing countries like China, India, Brazil etc., where a reasonable level of science exists, dependence on borrowed technology would be equally unwise. There are two major reasons for this argument. The first is that the location specificity of agriculture precludes the solution of a problem by an off-the-shelf product. Modification and processing of techniques for production of a relevant product in the agricultural sector requires indigenous capabilities for which both working knowledge and tradition of science are equally necessary. A delayed start of research in areas like molecular biology, which are basic to biotechnology and genetic engineering, would impose a very heavy opportunity cost which should be avoided by deliberate action. Secondly, a very large proportion of materials and methodologies, for crop improvement through genetic engineering, are being developed abroad in its private industry domain. For this reason, they are not likely to be easily, and readily, available to the developing countries at prices that they can afford.

6.2 Concerns

The advent of genetic engineering and biotechnology has already generated some concerns which have become points of vigorous debate in the industrialised world. The developing countries would do well to take note of these and draw appropriate lessons. The more relevant of these concerns are:

● *substitution of some traditional export commodities of the developing countries by biotechnology products*, a good example of which is the fructose syrup for cane sugar. It must be realised that, other than building world opinion on humanitarian considerations, there is very little that can be done to avert the problem. For the developing countries, the option is really to develop methods of technology forecasting and to build indigenous strengths to protect their interest

● *the new technologies, being developed primarily by private industries and multinationals, are protected under patents and intellectual property rights.* The patented material is increasingly including natural genetic resources. The germplasm in future, it is feared, will not flow as freely as it should. Already the debate on whether the developing countries, which are the major source of germplasm, should not be compensated is assuming worrying dimensions

● *biosafety of the field releases of genetically engineered organisms* has become a major issue in the industrialised world and is bound to have repercussions in the developing countries. The developing countries must initiate early action for the enactment of legal provisions, and other regulations, to ensure safeguards against any possible adverse effects

● there are also some concerns which will affect the developing countries exclusively and deserve a thorough debate to identify

547

optimal solutions. The major ones are

○ The possibility of the political colonisation of the past extending into scientific colonisation of the future. This is a sensitive issue and is likely to evoke strong emotional reaction in both the developing and developed world. But the issue is a real one and sweeping it under the carpet is likely to aggravate, rather than minimise its impact. It will be sensible to initiate debate and dialogue for finding a solution, and for generating educative information so that decision making on scientific issues is not confused by emotional reactions

○ The gap in the standard of science and research capabilities between the developed and developing countries is another concern,

○ Unfortunately, the gap is getting wider rather than narrowing. The developing countries will have to find their own solutions to this problem. Larger allocations on science and education are an obvious necessity. Equally important, however, is the administration of science and the place given to the scientists in society. Commenting on the reasons why science and technology has lagged behind in the South, Abdus Salam, President of the Third World Academy of Sciences, has said the following about the manner in which the enterprise of science has been run in countries of the South: *"An active enterprise of science must be run by scientist themselves and not by bureaucrats or by those scientists who may have been active once but have since ossified"*. In the developing countries, in general, the malady of science administration is a very serious one and Salam's prescription should go a long way in ameliorating the malady.

6.3   Strategy

As stated earlier, the development of a strategy for biotechnology application in crop improvement will depend upon the nature of problems faced by a country and its status of scientific development.

Two things, however, are clear : (a) the application of biotechnology has to be warranted by its need, and not because a tool is available that must be applied, and (b) a close integration of genetic engineering and biotechnology with conventional crop improvement research is imperative. The integration has to be deliberate rather than opportunistic.

The implementation of the strategy will involve : (a) identification of an economically relevant problem for which an efficient solution through conventional means is not available, (b) development of infrastructural facilities for doing molecular biology research, in general, and strategic problem solving research in particular, (c) a conscious effort to build trained manpower by a special drive within the country, and by creating opportunities for training in developed countries abroad, (d) developing international collaborative programmes with interested laboratories in the developed countries. Of particular relevance would be collaborative programmes and cooperative activities between developing countries. The advantage would be the similarities of problems that need solutions, and the common use of the developed products.

6.4 After Biotechnology: Raising Yield Ceilings and Identifying Untapped Reservoirs

There is now a general impression that the yield levels achieved in wheat, rice, sorghum and millets in the early seventies have not been surpassed. Is it a national perspective or an international phenomenon? Since the yield in the farmers' fields is dependent on many variables, including resource input, environmental factors and social aspects, an answer to the question can only be found through experimental trials in the All India Coordinated Projects. In wheat, URT are conducted in various regions. At Ludhiana, the maximum yield of a variety of 64.7 q ha^{-1} was obtained in 1977-78. After that there is a declining trend. In last four years the maximum of the best variety was 48.7 (1983-84), 47.1 (1984-85), 35.9 (1985-86) and 37.1 (1986-87). A similar, but not to the same extent, trend can be observed at Pantnagar and Hissar. This trend is a cause for concern since we are looking at a major crop such as wheat. The yield of hybrids CSH-6 and CHS-9 of sorghum released almost a decade ago have also not been surpassed.

Studies on wheat have clearly shown that the total dry matter production has not been enhanced and the increase in productivity is essentially due to an increase in the harvest index. There is also evidence to show that, possibly, the upper limit of harvest index has already been reached. Consequently, an improvement in dry matter accumulation, in combination with the present level of harvest index, could lead in raising the yield ceiling.

In India, temperature plays a very vital role in bioproductivity. Even C_4 crops such as maize and sorghum produce as much dry matter as wheat in winter. In the monsoon season the minimum temperatures are usually around 25°C, which is not conducive for net assimilation. This situation may aggravate if temperatures rise further due to the 'greenhouse' effect. Therefore, an important aspect should be to look for genotypes for higher net assimilation rate in relation to higher temperatures, plant nutrition and water consumption. The genotypes having advantageous traits in respect of the above could prove useful for increasing biomass and, eventually, raising yield ceiling.

7   Integrated Pest Management

Every year, agricultural produce worth Rs.6,000 crores is lost in India due to pests and diseases. Chemical control of pests, with pesticides, is the widely adopted remedy. In 1975-76, 55,814 tonnes of insecticides and fungicides were used in India and the figure rose to 1,44,625 tonnes in 1983-84.

The increased and indiscriminate use of insecticides has resulted in resistance, immunity and resurgence in insects. The use of synthetic pyrethroids, in recent years, has caused a severe whitefly outbreak in cotton in Gujarat, Andhra Pradesh and in parts of Tamil Nadu. Persistent residues of DDT and HCH, the chlorinated hydrocarbon pesticides, highly poisonous to human beings, have been found in vegetables, milk, oil, butter and meat as well as in the mother's milk.

The use of plant protection chemicals has, therefore, to be reduced. This can be done by developing an integrated pest management package which includes the use of biological control of pests. Biological control is inexpensive and, as a long-term control measure, causes no pollution, and poses no risk to human health. Biological control agents are abundantly available in nature. Several pathogens, including viruses such as nuclear polyhedrosis viruses (NPV) and granulosis viruses (GV), bacteria like *Bacillus thuringiensis* (B.t.), fungi like *Metarhizium Beauveria* and *Verticillium* and protozoa like *Schizogregarine* cause diseases in insects and destroy them.

Similarly, several insect-parasitoides (parasities thriving on insects) are also known in nature. Tricho-gramma is an egg parasitoid of several pests, in particular of the sugarcane internode borer. *Goniozus, Elastimus, Eriborus, Bracon, Trichospilus, Tetrastichus* and *Chelonus* are some of the other parasitoids attacking insect pests.

Insect predators are also commonly found. The coccinellid predators, *Chiloconus, Pharoscymnus, Cryptolaemus, Scymnus* and *Menochilus* feed on mealy bugs; coccids, scales and mites on citrus, grapevine and guava. The mirid predator *Cyrtorhinus* attacks the brown planthopper and the green leafhopper of rice. The green lace wing *Chrysopa,* another very effective insect predator, feeds on aphids, mealy bugs and young caterpillars.

Spiders also help in insect control. The wolf spider, *Lycosa,* is important in the rice ecosystem. Several insects and pathogens attack weeds and some of them have been developed as biocontrol agents.

The beneficial role of natural enemies, and the practical methods of using them under field conditions, have been investigated and exploited on a large scale in India during the last three decades.

The establishment of the Commonwealth Institute of Biological Control (CIBC) Laboratory at Bangalore in 1957 gave a fillip to biological control research. It introduced exotic natural enemies and trained entromologists. The Central Biological Control Stations set up by the Indian Council of Agricultural Research (ICAR), extended and intensified research on biological control at the national and regional level, covering a number of crops and weeds.

For the first time in India, private commercial biocontrol research laboratories have been set up and these include the Biocontrol Research Laboratory of Pest Control (India) Ltd., in Bangalore and the Main Biocontrol Laboratory of the Tamil Nadu Cooperative Sugar Federation, Chinglepet. The Bangalore laboratory is mass-culturing native parasites and predators and supplying them to the farmers, at cost price, for control of grape mealy bugs, coconut leaf caterpillar, sugarcane borers and other pests. The Chinglepet laboratory supplies the egg parasite *Trichogramma*, for the control of sugarcane borers, to growers in Tamil Nadu.

The Union Department of Biotechnology has come out with a scheme of setting up pilot plants for mass production of *Trichogramma*, predator *Chrysopa*, nuclear polyhedrosis viruses (NPV) for *Spodoptera Irura* (tobacco caterpillar) and *Heliothis armigera* (American bollworm) and *Granulosis* virus for sugarcane borer in Tamil Nadu, to cover 5,000 hectares through the Tamil Nadu Agricultural University, Coimbatore. Similar schemes are operating in seven other centres.

8   Livestock

We will now describe a proposal which, in its first phase, can help increase the export of Indian meat, which accounts for Rs 67 crores a year today, to Rs 500 crores a year in a five-year period. We have examined the possibilities and we find that Indian meat export can earn substantial amount of foreign exchange if action is taken to support this industry through (a) the creation of Island belts of rinderpest and FMD free zones in districts of high animal density, (b) the establishment of a modern abattoir to produce clean and hygienic meat meeting international meat standards, (c) the initiation of support programmes to provide inputs to a farmer to produce quality meat from buffaloes, sheep, goats and poultry and (d) the political and commercial will to combat bad publicity of Indian meat in international markets. This proposal will not, however, affect the population dynamics, internal Indian meat markets or prices.

8.1   The World Scene

The meat industry, although not much developed in India, is the most important food industry in the world. The total world meat exports per annum during the period 1980-82 amounted to US $ 21.09 billion, exceeding the contemporary exports of wheat and flour ($ 19.92 billion) and of milk and milk products ($ 12.72 billion).

Globally, meat production, which was 104.09 million metric tonnes in 1970, reached 1,46.90 million metric tonnes in 1985, registering an average annual increase of about 2.75 per cent. In 1970, the largest contributor was beef and buffalo meat was the (38.66 per cent) followed by pig meat (35.68 per cent) and poultry meat (16.98 per cent). However, by 1985, pig meat became the largest contributor (38.46 per cent) followed by beef and buffalo meat (32.13 per cent) and poultry (21.09 per cent). In general, the share of pig and poultry meat is on the increase in the world market.

8.2   The Indian Scene

8.2a   Statistical Summary

According to the 1982 Livestock census, the number of different species of livestock was *192.4 million cattle, 69.7 million buffaloes, 95.2 million goats, 48.7 million sheep. 10.7 million pigs, 3.2 million other livestock and 207.7 million poultry.* These livestock species contributed Rs 12552 crores to the national output in terms of milk, meat, wool, eggs and other livestock products and by-products in 1982-83 (National Accounts Statistics, 1985). If the drought animal power (DAP) contribution of Rs 6000 crores to agriculture and allied activities is also added to livestock, the total contribution to the national output comes to Rs 18,552 crores. During the same year the contribution from agriculture other than livestock was Rs 51,060 crores. The corresponding national output was Rs 1,62,895 crores.

Thus agriculture contributed 31.35 per cent to national output and livestock contributed to 7.7%. If agriculture and livestock are taken together, they contributed 39% to the total output. If the contribution of livestock in terms of drought power is deleted from agriculture and added to the livestock component, the relative share of agriculture and livestock to national output moves to 27.67 per cent and 11.38 per cent, respectively.

The plan outlay, during the Sixth Plan was Rs 97500 crores, of which agriculture accounted for Rs 12,539 crores (12.86% of total outlay). The outlay for dairying and animal husbandry was respectively Rs 460.3 crores and Rs 389.3 crores totalling to Rs 849.9 crores. Thus the percentage outlay for dairying and animal husbandry was 6.78 per cent of the agricultural outlay. These figures of outlay and contributions, revealing that the proportional outlay for dairying and animal husbandry is much less, suggest that there is a strong case for a higher plan allocation to livestock on the basis of investment-pay off criteria.

Indeed, to achieve an equitable distribution of income to the rural poor masses, the landless labour and the marginal farmers owning livestock, it is essential to make a higher budget allocation to the livestock sector. This allocation will also generate tremendous employment opportunities to people of rural India, particularly farm women, the prime target of present day planning.

8.2b   Milk Production

The enhancement programmes of milk production of the Government of India, the State Governments, and now the Technology Mission on dairying, are fairly elaborate. Substantial progress has been made, and milk yield has risen from 240 to 460 lakh tons a year. At present there are 80 lakh crossbred cows and 80 lakh more are in the pipeline. This single force will, at the end of the century, account for a large proportion of milk yield and leading to a milk glut for which we are not yet ready. We have therefore to plan now so that we can use this to our advantage. Our R&D arm is still too weak to provide the nutritional and health production support for this large production base.

8.2c   Meat Production

Meat production in India is characterised by low yielding non-descriptive buffaloes, goat and sheep raised primarily by millions of small producers with little or no land holding. Buffaloes form the mainstream of meat export, except for sheep and goat which are sold live or chilled to middle east. These animals are raised on crop residues, natural garbage and tree leaves. There is no organised meat trade in the country either for export or for distribution within the country. The slaughter facilities are non-existent and hygienic meat, and its processed products, are not available.

Two trends have characteristically come out since 1980, when buffalo frozen deboned meat was first exported from India. Between 1980 to 1988, the total foreign exchange earnings on buffalo meat alone have gone up to Rs 67 crores. This is contributed by small and marginal farmers who own buffaloes. These buffaloes are sold to metropolitan cities for milk production and the male calves are forcibly killed in order to save the cost on feed in the khatals. As per the present estimates, Delhi, Bombay, Calcutta and Madras, the four major metropolitan cities, annually contribute about 1 to 2 lakh calves which are killed between one to three months of age. The Government has made several attempts to launch schemes to salvage this precious germplasm without, so far, any success. The objective is to salvage these buffalo calves from the metropolitan cities and bring them back to rural farm families where the males can be raised to a weight of 100 kg (and then sold as buffalo veal, which in turn, can be exported). The female calves so salvaged can belong to the nation's network of milk yield. Farm families, too, will be able to earn extra income and employment will thus be generated both directly and indirectly. Finally, the export of meat will enhance the foreign exchange earning of the country.

This operation will depend on two major factors (a) the creating a belt where the buffalo male calf will be raised free from rinderpest and FWD, so that frozen meat from these areas will be acceptable even to EEC countries where the standards of import are highest and most rigid in the world (b) the development of slaughter and meat processing facilities satisfying the standards of FDA of USA and the EEC. In respect of organisms like salmonella and other VPH standards, this can be done through the development of modern meat processing complexes.

If a concerted effort is made on animal disease control, for which IVRI has developed the necessary technologies, and is in a position to produce seven vaccines in sufficient quantities, we can meet the targets of the Mission within a period of 3-5 years.

The objective of the technology mission would be to accelerate the pace of increasing rural income and employment primarily through buffalo development. But the mission would supplement efforts in all the other sectors of animal production as well. It will also accelerate the pace of application and the adoption of modern technology, e.g. safe and potent vaccines and embryo transfer technology, to improve productivity. The central pivot of the mission will be its large and effective meat processing units. Each unit will initially cost Rs 26 crores, out of which Rs 10 crores will be funded by the export meat industry, to make available target export earnings of Rs 36 crores a year beginning from the third year. The multiplier effect is expected to enhance the target, which may then be upgraded every year. The result would be that, effective from the third year, each additional district based unit would add Rs 36 crores to the foreign exchange. This way, in a period of 10 years, the foreign exchange earnings from buffalo meat alone could exceed Rs 500 crores. Some of the broad objectives of the technology mission are indicated below:

● Start with one district as a major model and involve 1,50,000 farm holds to raise 3,00,000 young buffalo calves salvaged from metropolitan cities.

● Develop a meat processing complex with the assistance of private meat export trade.

● Develop an organisational structure to service about 1.5 lakh families in terms of inputs like purchase of young calves, supply of concentrate feed of high quality and animal husbandry, and veterinary inputs at the door.

● Immunise all animals (buffaloes, cows, cattle, sheep and goat against FMD, HS and BQ). This will be the key sector to make the district free of all animal diseases which are responsible to reduce the cost of meat and make it difficult to export.

● Improving the average annual income of the participating farm families by Rs 250 per month. As the increase in the export acceptability and earnings increase, the pricing structure will increase the income to Rs 500 a month at the end of five years. The mission will also establish a system of monitoring the parameters of growth indicated above.

9   The Strategies

9.1   Sustainable Development Systems for Small Farmers

The concept of a small farm varies from country to country. In India, a small farm is one which is below 2 ha in size. Those cultivating farms below 1 ha in size are classified as *marginal farmers*. The cultivable land per capita declined in India from 0.48 ha in 1951 to 0.26 in 1981. It is expected to be 0.15 ha in 2000. In China, the per capita land availability is 0.11 ha. In the semi-arid, rain-fed areas of developing countries, small farmers may often be cultivating 10 to 20 ha of land but the yields are low and uncertain. FAO's *Study of Agriculture at 2000* shows that in most developing countries the average size of an operational holding will continue to decline. Besides the growth in population, the other factors contributing to this trend are (a) the withdrawal of farm land for industrial, communication, housing and other uses, and (b) inadequate opportunities for employment in the non-farm sector. Between 60-90% of the population in most developing countries live in rural areas and, for them, land and water based occupations, consisting of crop and animal husbandry, fisheries and forestry constitute both a way of life and a means to livelihood security. If the current global trends continue, we will witness, at the end of this century, two contrasting systems of agriculture. In one system, farmers will disappear and will be replaced by large agri-business companies. In the other, millions of families operating small holdings will be struggling to improve the productivity, profitability, stability and the sustainability of the farming systems. How can they be helped?

Before specific methods of helping small farmers are discussed, it would be useful to define the concept of sustainability.

Measurement of Sustainability

The core of this concept lies in equal and concurrent attention to intra-generational and inter-generational equity.

Intra-generational equity, which aims at giving a fair deal to the socially and econimically disadvantaged groups, involves attention to (a) issues such as adding a dimension of resource neutrality to scale neutrality, in technology development and transfer, thereby helping all farmers irrespective of the size of their holdings and (b) the innate input mobilising and risk-taking capacity to derive economic benefit from new technologies.

Inter-generational equity involves attention to the criteria for measuring the impact of irrigation and crop intensification on the water resources of an area, such as the rate of exhaustion of the aquifer, increase in salinity, etc. Inter-generational equity also involves attention to the protection of the productive base of agriculture. This implies the use of the basic agricultural assets of land, water, flora, fauna and atmosphere in such a manner that today's progress does not ruin tomorrow's prospects. Thus, sustainability has both economic and ecological dimensions. What are the operational implications?

The first requirement is a new definition of productivity. At a meeting organised at the International Rice Research Institute in December 1987, the following definition was agreed to for the purpose of integrating economic and ecological sustainability in one measurable equation:

Productivity = (output value/input value) + changes in environmental capital stocks

The value of output and input can be calculated with precision. We do not yet have readily usable methods of measuring, with reasonable accuracy, the impact on environmental capital stocks. At least three areas need measurement tools.

● *Soil Health* What is the impact of a given level productivity on the biological potential of land? Soil health is the summation of the physical, chemical, microbiological and erodability characteristics of the soil. Can we develop a reliable, speedy and low-cost method of measuring the biological potential of land?

● *Water quality and availability* We need suitable indices to measure the contribution of water to productivity

● *Biological diversity* What is the impact of new technology on genetic erosion both in crops and farm animals? Again, can we develop methods of giving expression to the impact of the spread of new technologies on biological diversity?

It is obvious that an urgent need is the standardisation of internationally agreed methods of measuring and expressing productivity in terms of its impact on the environmental assets on which sustained agricultural advance depends. This area of methodology development merits adequate technical and financial support. The emphasis should, hoever, be on simple tools of measurement which farmers can readily understand and use themselves.

Measuring sustainability by estimating the stability of productivity over time is best done taking an entire village or a water shed or the command area of an irrigation project or the catchment area of a river system as the spatial unit for measurement.

9.2   Support Services

9.2a   *Package of Technology*

To promote agricultural production of an economically and ecologically sustainable basis, under conditions of small farm holdings, *we must re-orient agricultural research to yield enhancement.* Yield enhancement depends on the genetic efficiency of the crop plant and the optimal management of the farm resources. Some areas of research which merit added attention are briefly described below:

Land and Water Management

Scientific land and water management holds the key to sustainable agriculture. As mentioned earlier, we need simple methods of measuring the biological potential of land. Agronomic and agricultural engineering research must help the watershed community to get degraded lands upgraded through appropriate land management systems, including agro-forestry systems. Social engineering research must help to persuade the farmer to review his methods of soil conservation and soil health care. There is still inadequate attention paid to the social engineering aspects of soil health management.

Crop Management

This will involve particular attention to *breeding varieties* which can optimise yield, under a given moisture regime, and to *feeding plants*, in a manner that the genetic potential for yield of a new strain can be fully realised. The breeding strategy in particular needs re-orientation. *We need to initiate a selection strategy where the F2, F3 and subsequent generations are grown alternately under conditions of moisture sufficiency and deficiency.* Such selections under divergent conditions of moisture stress, will help to breed varieties with the capacity to give high yields, when moisture is adequate, and reasonable yields, when drought conditions prevail. The *crop feeding* strategy should involve the use of legumes in the rotation (including stem nodulation, green manure crops), organic recycling and balanced fertilizer application in accordance with the requirement of macro- and micro nutrients in different soil types. A third aspect of crop management relates to the management of weeds, pests and pathogens. In addition, cultural and cropping practices, including date, density and manner of sowing and growing two or more crops together (in such a manner that the companion crops extract moisture and nutrients from different depths of the rhinzosphere) will help to reduce and distribute risks, and enhance yield and income. Risk aversion and distribution should be important goals of agronomic research particularly in dryland areas, since small farmers do ñot always have access to protective irrigation.

Energy Management

Energy management relates to the energy needed (a) for land preparation and crop management including sowing and harvesting and (b) for nutrient supply. In irrigated areas, energy management, particularly for pumping ground water or distributing conserved water, holds the key to the success of farming in years of drought.

Post-harvest Management

Besides the handling of the part of direct human value, such as grain, efficient post-harvest management implies detailed attention to every component of the biomass, with particular stress on the preparation of value added products from each of the components. Recent advances in bio-processing, microbiological enrichment of cellulosic material and chemical engineering have opened up new opportunities in biomass utilisation.

Farming Systems Management

In the ultimate analysis, the packaging of the best available technologies into an economically and ecologically desirable farming system, should be the end-point of all scientific effort. For this purpose, it is necessary to promote the optimum use of the available resources: climate, land, water, labour and credit. To achieve this goal, a two-pronged strategy is needed.

First, efforts are needed to achieve appropriate blends of traditional and frontier technologies. The frontier technologies which merit particular attention are : biotechnology, including both molecular biology and bioprocessing, satellite imagery, computer technology and microelectronics.

Second, the design of the farming system, whether it is wholly based on crops, or on crops and animals, or on crops, animal husbandry, fisheries and forestry - should give explicit consideration to promoting better linkages among the primary (agricultural), secondary (industrial) and tertiary (services) sectors of economic activity in rural areas. *Farming system research (FSR) specifically designed to promote better growth linkages is a must if we are to add the dimensions of income and employment generation in projects designed to increase food production.* We have pleaded for the conversion of the FSR approach into a Rural Systems Research (RSR) approach for this purpose (Swaminathan 1988).

9.2b   Package of Services

Knowledge and skill transfer, credit supply and the timely availability of the inputs needed for higher productivity are areas which have been receiving attention in our country since 1947. We, however, need more research on delivery systems with particular reference to the following:

● Synchronisation in space and time between knowledge transfer and the inputs needed to apply that knowledge.

● Synchronisation in space and time between credit supply and the inputs for which the credit is intended.

● Organisation of the services sector in the villages, e.g. seed production and distribution, integrated pest management and improved post-harvest operations, particularly by rural women and youth. For this purpose it will be necessary to train women in seed technology, integrated pest management and improved pest-harvest operation, and help them to organise effective service organisations.

● Helping to organise producer-oriented marketing arrangements and assisting farmers to avoid distress sales.

● Promoting, through training and appropriate infrastructure support, the growth of industries designed to produce value-added products from every part of the agricultural biomass available in the village (or in a group of adjoining villages). *In general, research on delivery systems which enable small farmers to derive full economic benefit from new technologies has not received the attention it deserves.* Such research should aim at drawing lessons from successes and failures in the field of technology transfer and help to design self-replicating systems of technology diffusion. Similarly, training procedures should be largely based on learning-by-doing, particularly where illiteracy is still widespread. Adequate attention to software development for the use of mass media like television and ratio is also essential, if the investment on the hardware components of such media is to confer the anticipated benefit.

9.2c   *Package of Public Policies*

The public policy package should include incentives for group cooperation in areas such as land and water management, integrated pest management, integrated nutrient supply, improved post-harvest technology and organisation of local level associations for sustainable development. In addition to reorienting extension services on a community basis, suitable incentives such as group insurance, group credit and group awards for community endeavour, in the conservation of environmental assets, will be helpful.

Besides public policy instruments for promoting this community endeavour and the adoption of knowledge-incentive, and ecologically desirable, technologies, an important, but often neglected, area of policy is the promotion of activities in areas where ecological linkages over widely separated areas determine a common agricultural future. For example, the severe floods in parts of north India and Assam witnessed almost year after year have highlighted the urgency of restoring damaged ecosystems in the Himalayas. Fortunately, due to an abundance of rainfall, much of the damage is still reversible. Unless mountain eco-systems are conserved, the agriculture of the plains downstream will be in jeopardy. In such areas, coordinated action may be necessary not only among different states in a large country like India, but also among nations all along the Himalayas. The same is true in the Andean region in Latin America.

Public policy should also foster the desirable form of research effort. As stressed earlier, an area of research relevant to sustainable agriculture is the integration of traditional and frontier technologies. With the powerful tools available today, agricultural technology can help to integrate scale and resource neutralities in technology development and transfer, and help to generate skilled jobs in the rural sector. They can also help to generate technologies for women which can help to reduce drudgery and introduce the twin concepts of flexi-time and flexi-location in relation to job opportunities to farm women.

Some emerging technologies which need particular attention from the point of view of integration with traditional technologies are:

- Molecular genetics and recombinant DNA technology
- Tissue and cell culture
- Sero-diagnostic techniques and vaccines for animal diseases
- Bio-processing and preparation of value-added product from agricultural biomass
- Microbiological upgrading of cellulosic material.
- Satellite imagery and remote sensing
- Computer Technology
- Micro-electronics.

In this context, we should make a distinction between science and technology. *Scientific research* leading to new knowledge and material can be generated in research institutions and universities. *Technology Development* is, however, best done in the farmers' fields, jointly by scientists and farmers. The failure to make this distinction often leads to difficulties in the diffusion of new methods and approaches, either because of economic and/or, ecological factors. Technology development should become a joint sector activity between scientists and farmers. We need urgently centres for technology blending and dissemination, which are inspired by the goal of bringing the best in modern science to the service of the small farm families. In this task, national organisations can also play a key role.

9.3   Agriculture for Exports

About one-third of India's GNP originates from the agricultural sector. However, agriculture's contribution to the total exports of the country is around 25-26 per cent. Agricultural exports are extremely attractive from the point of view of generation of factor incomes in the country (directly and through backward linkages). Per rupee of export to agricultural products, a factor income of 71 paise is generated at home, whereas the corresponding figures for manufacture and mining are 59 and 69 paise respectively.

If one examines the problem in terms of domestic resource cost, one finds that the cost of each rupee of foreign exchange earned is lower in agriculture than in manufacture. Given this perspective, it may be useful to look at agriculture as a potential foreign exchange earner.

Traditionally, tea, coffee, cashew, tobacco and spices have accounted for 55 to 60% of India's agro-based exports. However, in the years to come, the growth prospects in the bulk exports of these items are going to be limited by the emerging competition in the world market. It would therefore be necessary to either produce more sophisticated and value-added products in these areas or diversify into other areas of agricultural produce where the potential is vast and India's market share is low. In this context, two categories of products are worth exploring.

● Where export surpluses are already emerging.

● Where exports on a value-added basis can become attractive. It should be remembered here that India does not have the preconditions which are considered to be favourable for agri exports, viz:

○ Medium/low population density
○ Free cultivable land
○ Low domestic base as well as inelasticity of local consumption.

So far as the option of exporting value-added agriculture products is concerned, one has to examine the concept of technology-based agricultural export. There are several areas where such a potential exists, viz:

- Shrimp aquaculture
- Mushroom
- Processed foods

While exploiting the potential for agro export, the question has often been raised whether the domestic consumer will not be deprived through such an effort. What is not widely recognised is that the average productivity of a large number of exportable items, which have vast international potential, is very low in India. Should export ventures be set up in the areas of such crops, the technology for improving productivity will percolate down to the local farmer. This, in its turn, will also improve domestic productivity.

We believe that India not only has a potential for increased production, through more intensive cultivation of fruits and vegetables; but also that, being a country of continental proportions, having the advantage of lagged seasons in different parts of the country, India also has the competitive advantage in export marketing. If exploited adequately, it can result in India having a range of products available for exports right through the year, with the following advantages:

- Spreading the marketing overheads
- Reducing the vulnerability to specific crop failure in one region
- Ensuring a better presence in overseas markets.

In a centrally planned economy like ours, a clear cut Government policy will be necessary, which sets out the quantum and composition of export. A separate administrative set up for export will also be necessary so that we can avoid repeating mistakes that stood in the way of the export of industrial goods. Our past experience show that the chief lacuna was the conflict between those who were charged with promoting export and those responsible for regulatory administration. Plans and procedures will have to be put into place which ensure smooth operation of a well-directed export policy.

9.4 Need for Anticipatory Research and a Compatible Research Set-Up

There is a constant shift, in qualitative terms, in the demand of agriculture produce as also of the production environment. A striking example of the former is the greater demand for white bread and products of animal origin, with the improving economic standard of living. (It has also been observed that great consciousness for health and physical appearance is turning many in western world to salads as meals). An example of the changing environment in which agriculture will have to operate in the future is the increased CO_2 concentration, and other greenhouse gases, leading to global warming of the planet and rising levels of the oceans. It is clear, therefore, that the agriculture of the future will have to operate under drastically changed conditions. Our ability to cope with changes will depend upon our preparedness in terms of developing newer materials and technologies. This will also entail that we consciously develop strengths in anticipatory research so that we can face the challenges of future. Even though it may take extra effort to justify our actions (and a departure from our preoccupations with immediate problems) the needs and demands will be inevitable and the planning process, and administrative procedures, must be alert to this imperative.

An essential prerequisite for this is to develop strong schools of basic research, and centres devoted to the pursuit of excellence. Admittedly, fundamental research is often considered an unaffordable luxury in our country and this has led to the neglect which is responsible for the ever widening gap in the knowledge base between the developed and the developing countries. It is our strong and considered view that a time has come when our entire thinking needs a drastic revision. It must be clearly understood that, far from being a dispensable luxury, fundamental research is a crucial tool for sustained progress of applied research. If this argument is correct, and if it is agreed that basic research of international scientific standard is to be given its rightful place, we will have to create conditions in which excellence can truly be achieved. Adequate and sustained funding over a reasonable time frame, functional autonomy to the scientific teams to carry out work without administrative and bureaucratic confines, the ability to assemble compatible groups of workers and the freedom of movement to keep abreast with scientific developments by personal contacts and attendance at international scientific meetings are some of the important factors that will need attention.

10   Recommendations

Land

The constantly shrinking size of farm holding and increasing demands of enhancing production per unit area require attention to rigorous land and water use planning. The land use pattern must pay adequate attention to soil conservation and fertility restoration, so that intensive use can be achieved in a sustainable manner. Land use plans must ensure that land is used, but not abused. A consensus must emerge on a minimal level of productivity from a unit holding so that fragmentation is not allowed to go unchecked.

The time has now come to seriously expolore the scientific use of land to ensure proper soil health care and the protection of soil fertility. This will require attack at two fronts.

● Attention to research needs, which include collection of information and mapping the status of soils in relation to their physical characteristics, nutrient status, organic matter and microbial flora, etc. For making these inventories on the basis of agro-ecological zones, we must use relevant and advanced research methodologies of remote sensing and others. The land use will also need monitoring on a regular basis.

● A district level land and water use board can be considered for making land use plans. Members of the Board may include scientists, farmers, developmental administrators, leading banks of the area and credit institutions.

Water

Water is the other essential ingredient of the agricultural production system. An integrated water use policy will require consideration of all forms of water, including river water, land water, underground aquifier, sea water, sewage and effluents. Conservation and scientific use of water, again, requires attack at two levels.

● The research front, which takes into consideration manageable catchments with due consideration to the soil type, percolation losses, conservation against evaporation losses and the use of chemical and physical agents as protectors. Linkages between agricultural research institutions and chemical and petroleum industries should consciously be forged for this purpose.

● At the level of planning, conservation and water use can be overseen by the planning boards suggested for land use. A major objective of the water use policy has to be to obtain at least a production of 4 tonnes of grain per hectare annually from irrigated holdings.

Biological Diversity

Biological diversity is the feed stock for developing productive genotypes. Biological diversity also deserves to be conserved in its own right as a responsibility to posterity. Integrated plans for conservation and use of biological diversity demand *ex situ* and *in situ* policies using *in vitro* and *in vivo* techniques. In view of predicted changes in global climate, and the consequent shifts in pattern of diversity at species and genotype levels, detailed studies, with biospheres as the units of investigation, could provide a valuable lead to management of biodiversity with conservation objectives. The flora and fauna will need in-depth study both at micro and macro levels.

Technology

While technologies emerge from good science, their validation and application are in the hands of the users. Therefore, unless technologies take the requirements of location specificity into account, their use will be frustrated. The enhancement of sustainability is also closely linked to location specific exploitation. The most practical way of relevant technology exploitation would be to forge joint scientist-farmer development centres.

Agricultural exports

In order to provide diversity and resilience to Indian agriculture, serious attention will have to be given to develop it as an industry with a potential for exports. To achieve this, market surveys and attention to requisite quality considerations is imperative. Two considerations deserving attention are the specific requirements of the target importing countries and the prioritisation of our capabilities to tackle gradations of sophistication. For instance, it may be relatively easy to cater to the Middle East and African markets, for the same monetary quantum of a given export, compared to a Western market for an alternative item.

This will need a detailed assessment of specific quality attributes, required in a targetted export, and critical market surveys of importing countries abroad. Agricultural attaches in major embassies abroad, who have the technical qualifications for a proper assessment of the situation, would be crucial for this purpose.

Human Resource Development for Rural Prosperity

An important requirement of agriculture today is to attract and retain professionally qualified youth in rural programmes. This is essential for (a) providing intellectually absorbing, and financially remunerative, jobs in the rural sector, (b) to promote the well being of remote and neglected areas, (c) to overcome regional imbalances, (d) to remove commodity imbalances, and (e) to create the environment for utilising first rate scientific work.

A conscious promotion of rural industries, based on the major agricultural produce and agricultural by-products, deserves special attention if we are to meet the challenge of jobs for rural youth. Also, the creation of value added products will produce the capital for building rural prosperity. For example, the promotion of pineapple and orange based canned fruit and juice industries in the North-East would save both the highly perishable horticulture produce and create avenues of gainful employment in a region of our country where means of transportation are inadequate and regional sensitivities high.

Knowledge-base

Agriculture today has progressed from an intuitive exercise of the farmer to a highly science-based activity. For sustained development, it will have to absorb even higher increments of hard science. The enlargement of the science base in agriculture, including inputs from basic sciences like mathematics, chemistry, physics and biology, will be critical. The development and nurture of strong schools of excellence in basic sciences relevant to agriculture such as genetics, physiology, biochemistry, microbiology, plant pathology, entomology and others are essential, and a part of the inevitable route to achieve this objective. The application of knowledge-base frontier technologies will play an increasingly important role in modernising agriculture into a competitive industry.

Restructuring Education and Research

A sound educational foundation is an essential prerequisite for high quality research. In India we adopted the land grant model of the United States for higher education in agriculture. This paid dividends in an early thrust, but all agricultural universities unfortunately have weak programmes in basic sciences. A basic knowledge of mathematics, chemistry, physics and biology is essential for developing strong foundations in most disciplines in agriculture; disciplines which constitute the backbone of biotechnology, which is expected to play a major role in agriculture. Therefore, it will be necessary to have a strong interaction between agricultural universities and conventional universities. The following mechanism is suggested:

- Joint ICAR-UGC Committee for developing core courses and for exchange between conventional and agricultural universities.
- Joint research programmes on the pattern of USDA and universities.
- Training in basic science subjects.

Climate Change

There is now substantial evidence that the greenhouse gases such as CO_2, methane, CFCs, nitrogen oxides etc. are increasing in the atmosphere. As a result of this, it is predicted that global climate will change, and we can expect a warmer global climate, coupled with uncertain precipitation. At present, the sources of CO_2 and CFCs emissions are known, but there is considerable debate and uncertainty about methane and nitrogen oxides. There are suggestions that considerable amounts of methane are produced from rice paddy and ruminant cattle. Studies are needed on the following aspects :

- Estimates of various greenhouse gases contributed by developing nations, particularly, India, China and other South-east Asian Countries.

- In view of the increase in temperature and the uncertainty in precipitation, we need to place emphasis on developing research programmes to meet the anticipated climate change; we must strengthen research on tolerance   to abiotic and biotic stresses in different crops and animal species.

- The coastal areas need special attention because the sea-level rise may damage coastal ecosystems.

Energy Balance in Agriculture

The agriculture sector contributes about 34 per cent to the total gross domestic product, but the share of agriculture in commercial energy consumption has remained around 2.5 per cent. If we consider the contribution of agriculture to the energy sector, particularly the non-commercial energy, it may be approximately one third. Thus agriculture is, at present, a source of energy to the total energy sector of the country.

However, if agriculture has to be modernised to meet the challenges referred to earlier, it would be essential that more energy is consumed by the agriculture sector. While doing so, we must ensure improvement in the efficiency of different inputs which need energy. This will require an analysis of energy balance in different crops and different farming systems. It would then be possible to identify the inputs where energy conservation is possible. In this context the role of draft animal power now, and in future, needs to be assessed.

Personnel Policies

Good science comes out of an imaginative system. In its absence the existence of talent, and good intentions, alone will not work. It is also generally agreed that the low ebb at which we find agriculture science and research today can be traced to the centrally structured bureaucratic organisation under which the system operates today. We must correct this situation immediately. But even as we effect the change, we have to be conscious that a totally foreign explant may find violent rejection. It is in this context that a careful assessment of the various alternatives experimented in the country will be prudent. The important elements of a workable system are (a) operational autonomy for assembling compatible groups of talented and motivated scientists, (b) a salary structure that can attract and retain the recruited talent, (c) adequate finances to meet the cost of scientific demands, (d) flexibility to utilise the financial resources for making scientific progress, (e) opportunities for being in the thick of the competitive international science scene, (f) an environment which respects the sensitivity of a creative mind, (g) the realisation that agricultural sciences cannot be pursued in isolation of the endeavours in the other disciplines of science.

It is perhaps because of this realisation, of the deficiencies of the existing system, that the Government has decided to embark upon an overhaul of the ICAR system, and bring agricultural research and education under the charge of the Prime Minister. This is a very welcome step and provides the opportunity to usher in the new concepts needed.

A critical assessment, based on the principles outlined above, suggests that an extension of the basic structure operative in the CSIR system to agriculture research will be most appropriate. Like in the CSIR, the Prime Minister would be the Chairman of the ICAR Society. An Advisory Board and a Technical Advisory Board can be constituted on the same considerations as in CSIR. Each Institute, can likewise have a Research Council consisting primarily of eminent scientists in the relevant field (with a scientist outside the system working as its Chairman) and other appropriate inputs (progressive farmers, industry representatives, social scientists, etc.). The recruitment of scientists and technical personnel like in the CSIR would be done at the Institute level at all levels other than the Director. The Directors can be identified by constituting search-cum-selection committees on the pattern of CSIR.

12
Chemical Industry in the year 2000 AD: National Priorities

The chemical industry can play a decisive role in alleviating India's chronic problems. Indeed, it can be the country's instrument of change, propelling it to a flourishing future.

Contents

12
Chemical Industry in the year 2000 AD: National Priorities

Preamble

The chemical industry plays an important role in meeting the basic human needs of food, shelter, clothing and health. The Indian chemical industry is unique due to a variety of reasons. These include the enormous growth of the nitrogenous fertilizer industry, lack of availability of free foreign exchange, inevitable emphasis on synthetic fibres in the future due to the pressure on land, acute problem of housing, etc. Although the chemical industry in India is barely thirty years old, it is poised for a big growth and the expected investment in this industry during the next 15 years is around Rs 70,000 crores, excluding the downstream processing industry for plastics, elastomers, fibres, etc.

The Indian priorities in the area of fertilizers, petrochemical processing, agrochemicals, bulk drugs, synthetic detergents, chlorine-based chemicals, speciality chemicals, polymers and composites and biotechnology have been highlighted. The question of renewable and non-renewable resources, as well as constraints restricting the growth of the industry, have been examined. Apart from highlighting the priorities in each of these areas, four specific areas pertaining to fertilizers, synthetic fibres, speciality chemicals and high performance polymers have been identified as areas where, with proper inputs and policy changes, India can assume a leading position.

More than 30 per cent of the investment in the chemical industry during the next two decades will be in the fertilizer industry and needs and opportunities in this sector will have to be examined closely. Methods to increase the efficiency of urea utilisation, through large granule or coated urea formulation, integration of urea flow sheet with soda ash plants, change-over from mega size ammonia plants to lower capacity plants, acidulation of phosphate rock with nitric acid to obviate the use of sulphuric acid, etc. should be the main priorities. Due to a variety of reasons, India can acquire a leading position in the fertilizer sector starting from supply of technology to commissioning of plants, provided this opportunity is seized properly.

By the turn of century, India may become the largest synthetic fibre producer (particularly of polyester fibres) in the world. There is a great export potential. India can indeed assume a leading position in this sector provided appropriate R&D inputs, with a focus on the development on new products with Indian needs in view, are put in.

In view of the existing structure of a well-organised small scale industry, and large trained manpower, India can make a major contribution in the area of speciality chemicals and intermediates as well as finished products, including the so called fine chemicals, high purity chemicals for electronics, etc. These could be produced for domestic consumption as well as for exports.

There are certain areas such as engineering plastics or high performance composites where technologies will not be available at any cost. Hence urgent and massive R&D inputs in these areas are recommended. ∎

1 Introduction

The chemical industry plays a unique role in meeting the basic human needs and desires. The problems of ensuring sufficient food, adequate shelter, reasonable clothing and minimal medical help for a population in a country like India, which may exceed one billion by the year 2000, are truly staggering.

The perspectives of the chemical industry in India are governed by factors considerably different from those in USA, Western Europe, Japan, etc. A few specific reasons can be cited. For instance, the nitrogenous fertilizer industry in India will have to necessarily grow in a big way unlike the situation in the developed countries, where this industry is already stagnant, if not on the decline. The availability of free foreign exchange is always limited. The pressure on land is severe and displacement of cotton crop with food crops may become increasingly important, leading to greater emphasis on synthetic fibres, particularly polyesters. The problem of housing is acute and the backlog runs to more than 30 million houses.

The chemical industry can play a decisive role in alleviating India's chronic problems and raising the standard of living; it can thus be a major instrument of change, propelling India to become a flourishing country.

2   Background

The chemical industry in India is a late starter and is barely thirty years old. Petroleum refineries were established in 1954. Initially the sugar industry provided a surplus of molasses, whose disposal used to be an acute problem. Through a policy decision of the Government of India, cheap ethyl alcohol, based on throwaway price molasses, provided the base for the synthetic organic chemical industry. Small capacity plants for polyethylene, polystyrene, synthetic rubber, acetic acid, etc., were installed from the late fifties to the late sixties. Most of the plants were based on technology acquired from overseas. Lately, in the generation of indigenous know-how, particularly for speciality chemicals required for dyestuffs, drug and pesticides industries, we have had some remarkable success. The prices of chemicals, plastics, synthetic fibres, synthetic rubber, etc. have been influenced by the high cost of raw materials and imposts of several levies and duties; the free market prices have been tampered by the import duty. Of late, the Government has enunciated the welcome policy of economic size of plants for basic chemicals, polymers, elastomers and synthetic fibres. The expected investment in the chemical industry during the next fifteen years is around Rs 70,000 crores, excluding downstream processing industry for plastics, elastomers, fibres, formulation of drugs and pesticides, etc.

As regards resources, a substantial amount of offshore and onshore oil and natural gas is available and is equivalent to about 40 million tpa of crude oil. There is no indigenous sulphur. The forest resources are limited and have to be assiduously preserved and expanded; the last three decades have witnessed systematic and devastating denudation of forests.

Although India's organic chemical industry was born on cheap fermentation alcohol as the feedstock, we can see, today, that this source will not be viable for various reasons, including a much higher cost and a lack of availability. Thus the future is inextricably connected with petroleum feed stocks. But this should not be a source of worry. We have got a considerable amount of ethane and propane in the natural gas, which can provide a valuable base for ethylene and propylene. The quantum of naphtha, which is required for making olefins as well as aromatics, should be available. In any case, the percentage of total hydrocarbons to be dedicated to basic petrochemicals will be less than 8 per cent (by contrast, for fertilizers it may be around 20 per cent).

3   Fertilizers

3.1   Nitrogenous and Phosphatic Fertilizers

More than 30 per cent of the investment in chemical industry during the next 15 to 20 years will be in the fertilizer sector. We will perhaps have a situation where out of all the hydrocarbons consumed in the country, the highest fraction will be earmarked for ammonia production (higher than any other location in the world). The agro-climatic conditions are such that we really need all types of nitrogenous fertilizers such as, urea, ammonium nitrate, calcium ammonium nitrate (CAN), diammonium phosphate (DAP), etc. The present policy of total conversion of ammonia to urea should be critically re-examined since urea is believed to be a good fertilizer only for anaerobic conditions and for clay type soils. Further the utilisation efficiency of urea by plants is really low and this gets aggravated by high ambient temperatures in India. We will need a multi-pronged attack.

India does not have any indigenous sulphur and there is a forecast of world shortage of sulphur, and an increase in prices in the 1990s. The acidulation of phosphate rock with nitric acid may prove to be a solution in India as it obviates the use of sulphuric acid which eventually ends up as calcium sulphate. Here the entire N is kept in the desired nutrient form and low grade phosphate rocks can be utilised. Elsewhere in the world, the compulsion to use nitric acid is not so great and thus such processes need to be upgraded in a major way. A careful investigation of the use of low grade phosphate rock, for manufacture of nitrophosphate as well as substitution of imported sulphur by indigenous iron pyrites, should be taken up on an urgent basis.

The adoption of the Solvay type process for calcium nitrate, via the use of unconverted ammonium carbonate, has probably not been practised anywhere but we can see the distinct advantages of this strategy in India. Further, this will allow liquid fertilizers to be sold in highly developed agricultural sectors, where a mixture of ammonium nitrate and urea is supplied, which ensures quick and slow release of N. However, we will have to consider the question of land holdings, farming practices, additional infrastructure requirement, problems of quality control, etc. before such a strategy can be accepted.

Recent reports indicate that urea-nitrophosphate fertilizer, which is sufficiently acidic to prevent volatilisation of ammonia, may be worthwhile considering. In this strategy, phosphate rock, after acidulation with nitric acid, is mixed with molten urea and subsequently dehydrated to give suitable grade materials. After a careful cost-benefit analysis and agronomic acceptability, this strategy can be employed advantageously.

583

In some plants in India there is a scope to integrate the urea flow sheet with the soda ash plants, thereby reducing capital investments in both the plants. In fact, due to large paddy cultivation, ammonium chloride obtained in the dual process for soda ash, can also be advantageously utilised, especially for water logged conditions for paddy.

There is an urgent need to increase the utilisation efficiency of urea. Strategies such as split dosing have been already recommended but the need to develop slow release urea either as large granule urea or coated urea with 2 per cent to 4 per cent formaldehyde is obvious. For the latter purpose, a large ammonia-urea plant can be integrated with a decent size methanol plant advantageously, and methanol can be converted to formaldehyde on-site.

The wisdom of putting mega size (that is, 1350 tpd) ammonia plants, and the associated urea plants, is now being questioned all over the world. The belief that the lower capital cost per annual tonne of ammonia will bestow definite benefits may prove to be a myth. The garnering of large capital requirements is also not easy. Thus flow sheets calling for lower capacity at about 450 tpd, possibly with lower pressure synthesis at about 80 atm, will require closer examination.

The primary reforming of natural gas or naphtha can be modified substantially to allow higher pressure operation with a high methane slip. This should cut down the subsequent compression cost. The use of enriched oxygen, obtained through pressure swing absorption or a membrane process, can possibly take care of the N:H balance. The use of higher temperature gas from the secondary reformer can constitute the heat transfer fluid for the primary reformer.

There is an urgent need to develop better performance low temperature shift conversion catalysts and, more particularly, ammonia synthesis catalyst which can work at lower temperatures. The design of ammonia reactors should be reconsidered with radial rather than axial flow of gas and indirect rather than quench cooling of the ammonia synthesis gas. Better catalysts to reduce the catalyst loading should be also considered.

3.2   Potassium Based Fertilizers

Although the common salt (NaCl) production, based on sea water, is quite high at levels exceeding 7 million tpa, we have practically no recovered KCl for fertilizer applications, and imports are at a level of more than 1 million tpa. The desirability of converting KCl to KNO_3 merits consideration particularly as K and N are in a useful form in the latter.

3.3   Future Strategy

A comment on India's role in technology development/export in the late nineties should be made at this juncture. India is capable of becoming a leading fertilizer manufacturer in the world, provided the requisite inputs in R & D are made. Indeed India can acquire a leading position covering all types of nitrogenous and phosphatic fertilizers: from preparation of project reports to supply of technology, erection of plants, to running of plants on a contract basis, at any location in the world. A large reservoir of trained personnel in various sectors, accustomed to working under adverse and hostile conditions, is a special asset. Developing countries in Africa, South East Asia, etc, can benefit from the Indian experience. Thus, from technology to running plants, India and China may well emerge as serious contenders on world scene. It is important for us to seize the opportunity to become a leader in the field of fertilizers.

4   Petroleum Processing

Firstly, the demand pattern for petroleum products in India is heavily slanted towards middle distillates (kerosene and diesel) and secondly, the feedstocks of Indian origin, notably Bombay High, are 'unusual' in their characteristics. The key features of Indian crudes are : (a) an unusually high wax content which causes problems in transportation and (b) the lighter 'cuts' (naphtha, kerosene) from Indian crudes are usually high in aromatics whereas the heavier portions (gas, oil, residue) are highly paraffinic.

The waxy nature of the crude oil poses special challenges such as (a) development of a suitable pour point depressant; here a thorough knowledge of crystal growth and its morphology will aid the development of 'tailor-made' additives suitable for our conditions (b) wax also gets deposited in the pipeline and prevention of this will require a deeper understanding of the underlying process (c) also, waxy crudes exhibit complex rheology.

The unusually high aromatic content of naphtha and kerosene brings with it the need for processing technologies not required outside India. For example, Bombay High naphtha needs dearomatisation for efficient use as a feedstock for the petrochemical and fertilizer industries. Kerosene fractions from Bombay High and Assam crudes also require dearomatisation. One particular challenge in the latter case is to develop an energy efficient solvent extraction process which would be an alternative to the 'ancient' liquid sulphur dioxide extraction Edeleanu process.

The desire to maximise middle distillates by fluid catalytic cracking or hydrocracking of Bombay High vacuum gas oils, usually high in n-paraffinic content, poses special problems in catalysis and chemical engineering. There are nine FCC operating units in India all based on imported technology and catalysts. The operations i.e. yields of all these leave considerable room for improvement.

Indian refineries now plan to incorporate hydrocrackers for middle distillates maximisation. Besides the development of our own catalyst for this purpose, there will be new challenges in the area of trickle bed hydrodynamics, especially pertaining to stagnent zones and runways. The scale-up information is largely in the hands of large multinational organisations and kept confidential. Given the desire to develop the hydrocracking technology 'in-house', a concerted effort will be required to generate the know-how.

5   Agrochemicals

In the recent past we have done reasonably well in developing cost effective and occasionally innovative technologies for a variety of pesticides, including Isoproturon (without using phosgene), Fenvalerate (synthetic pyrethyroid), Cypermethrin (synthetic pyrethyroid), Quinolphos, etc. However, there is a need to tackle the special problems associated with nematodes and rodents. The development of growth regulators (particularly for sugar and cotton), development of encapsulated forms of toxic pesticides, enabling safer handling and controlled release, etc. are priority areas.

The wastage of pesticides during spray is a major problem in India, which has implications both as a health hazard and as loss of precious material. Some innovative work on the design of spray equipment, as well as the development of additives to control and narrow down the drop size distribution, is needed so that the pesticides will reach the targeted area more effectively.

6   Bulk Drugs

The production of bulk drugs in India should get a major impetus not only because of the reasons connected with the diverse needs of the growing population but also because of the tremendous export potential that this sector provides. In particular drugs for tropical diseases like enteric fever, apart from pain killers, will have to grow in a major way. Even the production of vitamins (particularly A, C and E) will have to increase substantially. Vitamin A should be based on a wholly synthetic route.

Natural materials from plant and marine extracts are proving to be very interesting sources of entirely new drugs. A closer and sustained study of systems of traditional Indian medicine could pay rich dividends.

7   Synthetic Detergents

There is a chronic shortage of vegetable fats in India, and animal fat is a taboo. Thus the growth of the soaps and detergent industry, based on this source, is not possible. The problem is aggravated by the presence of hard water in most places. Thus we need to develop our own recipes of toilet soaps and synthetic detergents, which take into account the problem of hard water. The wisdom of using only the conventional dodecylbenzene sulphonate sodium salt-based synthetic detergents is open to question particularly for reasons cited above, and because of the inevitable increase, in the coming decades, in the use of polyesters and blends in India.

There is a great opportunity to adopt alpha olefin sulphonates (AOS) in a big way. The alpha olefins should be based on ethylene, whose potential availability is very good. There is a need to develop technologies for oligomerisation of ethylene which gives predominantly alpha olefins in the range of C_4 to C_{16}. The potential for alpha olefins (C_4-C_{20}) could be at least 400,000 tpa in India by 2001.

The large scale use of synthetic fibres would require synthetic fatty alcohol (C_{13}-C_{15} range), ethoxylates and their sulphates. Due to lack of fats, as pointed out above, and coconut oil being three to four times more expensive than in the rest of the world, the manufacture of C_{12}-C_{15} alcohols by the synthetic route is desirable. There is considerable scope to develop hydroformylation of higher alpha and branched olefins to obtain the above materials.

8   Chlorine Based Chemicals

There is a very good potential for chlorosolvents, like carbon tetrachloride and perchloroethylene, to be produced on a large scale, and India can consider exports of 2,50,000 tpa of these materials as the cost of chlorine is really low in the international context. This apparently paradoxical situation arises out of a peculiar situation, where, unlike the Western world, the loading of cost is on caustic soda and chlorine is usually an unwanted by-product. The textile industry, besides the paper and detergent industries, consumes a fairly large amount of caustic soda. Further, with the growth of large alumina plants, which are export oriented, additional imbalances will be created. India can convert this liability into an asset and additionally consider large-scale, port based bleaching powder plants for export markets. For large plants the total energy concept is expected to be adopted.

An important area for chlorine usage can be in the production of phosphoric acid from rock phosphate and hydrogen chloride. R&D efforts, on handling dry or semi-dry reactions and extraction with special solvents, will be needed. This effort can have a major impact on avoiding the sulphur import as well as on the utilisation of surplus chlorine.

9   Speciality Chemicals

Speciality chemicals, including the so-called fine chemicals, and the high purity chemicals for electronics have a bright future. Many of these chemicals can be produced on a small scale and, in view of our large trained manpower, and the existence of a well-organised small scale industry, we can expect major contributions to emerge from this sector. Intermediates, as well as finished products, for plastics, rubbers, fibres, pesticides, drugs, electronics, etc. will be produced in a major way in the country for domestic consumption and for exports.

10   Polymers and Composites

Polymers play a pervasive role in our everyday life ranging from highly specialised applications in high-tech areas to mundane mass produced products found in almost every home.

Polymers can be found in a variety of forms; some of the dominant ones are plastics, rubbers, fibres, sealants and adhesives. Each of these polymers can be classified into the commodity (high volume-low price) and the speciality (low volume-high price) variety. Whereas technologies are usually available for import in the commodity area, those in the speciality area are closely guarded and not available for several reasons, including the strategic ones.

10.1   Commodity Plastics

The image of the plastics industry in India is that of a supplier of buckets, pans, mugs, footwear, etc. There is a general lack of awareness of the vast performance potential of these versatile materials.

589

This can be inferred from the low per capita consumption of plastics in India, which is barely 0.5 kg as compared to the world average of 11 kg; even China and Brazil have markedly higher per capita consumption than India.

Most of the bulk thermoplastics were introduced in India rather late. Although some of the polymers like LDPE, HDPE, PVC etc. were introduced in India in the late fifties, PP and ABS were introduced only in the late seventies. LLDPE, however, is an exception since it will be introduced in India within a decade of its commercial entry in the world market. Over the past few decades, the use of commodity plastics has been accepted by the Indian users in routine applications.

The commodity thermoplastics will grow in a big way largely out of the local needs and the lack of wood, metals and alloys. Some areas of applications are particularly critical for India. Packaging of vegetables and fruits is an immensely important area and it is a pity that the gulf between the grower and the user is wide; margins are very large and the wastage is also very high. We need to develop cheap barrier films (simple or multilayer construction) which will revolutionise this area. Thus the cheapest material we can consider for the barrier film is perhaps polypropylene or polyethylene as the structural component modified with nylon as the barrier component.

In the field of commodity polymers there is considerable scope for replacing conventional materials like paper, glass, metal, etc. Besides introducing new materials, new processing techniques will have to be developed for effective utilisation of polymers. Recent developments such as multi-layer extrusion of films for barrier packaging, multi-layer blow moulding, profile extrusion with wood or aluminium, etc. have to be introduced in India. The area of barrier packaging, with multicomponent systems providing improved mechanical stability and permeability, offers enormous advantages due to the possible reduction of the layer thickness of the expensive material.

The problems of packaging large quantities of fertilisers, sugar, salt, vanaspati, petroleum products, etc. present unique opportunities for large scale use of commodity polymers in India. Consider, for

590

example, the problem of producing 20 million tonnes of urea by the year 2000 and then the question of packaging this in 50 kg PP or HDPE woven bags. Clearly, a development effort for packaging would be a key area to pursue in view of the enormous potential.

Among the new application areas in commodity plastics, the following are particularly pertinent to India:

- Industrial fabrics such as coated fabrics, filtration fabrics, etc.
- Civil engineering fabrics or geotextiles
- Performance textiles such as upholstery and insulation
- Agricultural applications.

The demand of commodity polymers in India has consistently outstripped the production. Large capacities of commodity plastics will have to be established, with plant sizes comparable to the international ones. Large scale expansion of polyethylene, polypropylene and polyvinyl chloride is urgently needed. The absorption of manufacturing technologies, particularly for plastics such as LLDPE, should be a major endeavour.

10.2   Engineering Plastics

The enormous potential for the large scale usage of engineering plastics can be appreciated with some simple examples. The engineering plastic polycarbonate, for instance, has excellent transparency and impact resistance, and its sheets are shatter proof. There are great opportunities for replacing glass by polycarbonate in street lights, sign boards in banks, post offices, signals, etc. Lightweight, but strong, polycarbonate helmets and light non-metallic polycarbonate riot shields for the police are other examples of large scale usage. Engineering plastics have a tremendous scope for replacing specific metallic parts in two wheelers, which are poised for a massive growth in India.

At present there are number of constraints on the use of engineering plastics in India. These include the slow progress in the areas of development of materials and processing machinery, the non-availability

591

of sophisticated processing machines (which could be microprocessor controlled and which will enable economic mass production of engineering parts), the lack of appropriate tool design and manufacture and the resistance to product acceptance in conventional industries.

It has to be emphasised that the engineering polymers that have already been introduced in the country such as PET, PBT, Nylon-6, ABS and (engineering) polypropylene account for more than 50 per cent of the engineering applications that have been proven abroad. Removing the constraints mentioned above would aid considerably in promoting the use of engineering plastics in country.

10.3 Speciality Plastics

Synthesis of speciality plastics (such as polyimides, PEEK,etc.) with exceptional thermal performance, polymers such as PVDF with piezoelectric and pyroelectric properties are examples of leaders in the field of speciality plastics. Indeed speciality polymeric systems that can effectively couple or convert thermal, electrical, light (solar), mechanical and perhaps other forms of energy should be developed increasingly. These will find extensive applications in high techn ogy areas such as in space and defence. Some of the applications will be so specialised that the technologies will not be readily available for import and India will have to have its own R&D efforts in these areas.

10.4 High performance composites

The era of advanced polymeric composites holds promise for revolutionary advances in structural materials. High performance polymer composites are made by using strong, stiff fibres (such as carbon, aramid, boron), in the desired orientation, which are then bound together with suitable thermoplastic or thermoset matrix materials. Advanced composites provide opportunities for development of new materials which will surpass the strength of structured metals and alloys on an equal weight basis.

The present combat aircraft use close to 30 per cent composites and this is likely to increase to about 50 per cent in the next generation combat aircraft. Presently, the use of such composites is growing in applications where performance, rather than cost, is the key criterion. Helmets, body armour, light weight segments of movable bridges, portable rocket launchers are some key examples. High performance composites (such as carbon fibre/epoxy) have great promise in the industrial machinery as reciprocating or rotating parts where the weight, strength, fatigue life, noise and natural vibrational frequency of metals limits the operation speeds and, thus, productivity.

In India, we cannot simply afford to lag behind in this high technology area considering the enormous impact it is likely to have in our space and defence capability.

10.5 Polymer Alloys

The technique of blending or alloying polymers could be effectively used to tailor-make materials, in a cost effective manner, for specific application needs. There is a tremendous scope for developing elastomer-elastomer, thermoplastic-elastomer and thermoplastic-thermoplastic blends. Polyphenylene oxide-polystyrene blends are outstanding examples of blends in extrusion, injection moulding and structural foam moulding applications.

The potential for development of polymer alloys is virtually untapped in India, although a number of constituent polymers of potentially attractive alloys are already available. For instance, ABS plastics are relatively very expensive in India, and innovative modifications of polystyrene, to give high impact polystyrene and modified polypropylene with higher impact strength (with EPDM, SBS, etc.), may lead to cost effective materials. The cost of ABS can be brought down through new alloying methods utilising PVC resins.

10.6 Fibres

10.6a Polyesters

For the Indian climate polyester and polyester-cotton (or viscose or polynosic) blends are ideal in almost every respect. However, there

593

are distinctive features of the Indian scenario. The market, in an organised way, for uniforms in school, police, military personnel, etc. is very large indeed. Further, the market for undergarments can be exceptionally high and hosiery type materials for a variety of uses have yet to be developed. A speciality in the Indian context is the *sari*, where there is a lot of scope for further improvements in the polyster fibre or filament. Due to high humidity in many parts of India, we would like to have filament and fibre with a comfortable feel and good water absorbancy. In the case of viscose fibres, improvements have been realised by incorporating a small amount of sodium carbonate before spinning; the release of carbon dioxide resulting due to such incorporation imparts the desired porosity. The comfort of PET can be improved either by chemically incorporating hydrophillic moiety into the polymer (which is an expensive route) or by generating an irregular rough fibre surface through spinning modifications, such as asymmetrical quenching of the filaments and non-circular cross section spinnerettes. Recent innovations relate to a clever manipulation of hydrophilicity and hydrophobicity coupled with a structure to promote maximum moisture migration. There is a considerable scope for R&D/product development with the Indian needs in view. Export potential in this sector is also high.

10.6b Polypropylene Fibres

This is the cheapest synthetic fibre with several advantages. We see an immense potential in India for non-woven fabrics based on polypropylene, including that for cheap blankets. The technology would involve air-laying processes for the fibres followed by mechanical or hydraulic needling. Thus aerodynamic studies have to be made in great detail. In order that the blankets be cost-effective, spin-bonding technology may have to be modified to produce a 'lofty' mat versus the conventional flat sheet.

Polypropylene fibres and filaments can make a major dent, on a cost-effective and fundamental basis, in linens, draperies, upholstery (particularly for public transportation services) etc.; leather cloth based on PVC is not very comfortable due to the hot humid climate. Here, new spinning and texturing methodologies will have to be developed. It may be emphasised that, due to the cost-effectiveness coupled with functional characteristics, polypropylene ropes for fishing nets and other purposes are extensively used in India in place of nylon ropes.

10.6c High Performance Fibres

India will need polyamide fibres (aramid type) for a variety of end uses; the bullet-proof vests for personnel in police, military, etc. will consume a fair amount. In these hightech areas, it becomes imperative to develop technology as, in all probability, the purchase of know-how will not be feasible.

10.7 Elastomers

The upgradation of natural rubber is necessary as India produces about 2,50,000 tpa and, for many applications in a tropical climate, it is highly desirable to modify the properties. Thus, epoxidation of natural rubber is a distinct possibility. Such types of reactions have so far received scant attention.

The canvas shoe market in India is unbelievably large, and the use of conventional rubbers, requiring vulcanisation of rubber, is not wise. The adoption of thermoplastic elastomers, based on styrene-butadiene-styrene triblock polymers, is highly desirable. Here the wastage of rubber is avoided.

The nonavailability of chloroprene calls for the development of alternative rubbers, which are oil-resistant. The recent development of combining ethylene-vinyl acetate copolymer with polyvinylidine chloride is worth emphasising. In certain cases of thermoplastics and rubbers some important property improvement can be realised through hydrogenation, using Raney nickel type catalysts.

10.8 Distinctive Applications of Polymers in Key Areas

The application of polymers can have an impact in some distinctive areas for a country like India: railways, bicycles, waste land development, housing, etc. are some examples.

The railway network is indeed gigantic in India and will certainly continue to grow by leaps and bounds. In fact, one just about never finds a vacant seat and compartments always 'overflow'. The role of plastics, composites, rubbers, etc. in the modernisation of the railways can be crucial. The use of materials like phenolic/PVC/polyurethane foams for wood replacement, and FRP sheets and pultruded forms for metal replacement in a railway coach would reduce the weight by about 30 per cent and correspondingly increase the payload. This will also reduce pilferage drastically as a lot of brass, and other easy-to-dispose materials, would no longer be available. This would involve innovative chemical formulations, processing technology and product design inputs. Take the case of signals alone: imagine the cost reduction that would be possible with composites and transparent plastics like polycarbonate, which will also make it somewhat less prone to attack by vandals. The role of elastomeric compounds in shock absorbers and in noise reduction can be quite substantial. The basic problems in these areas, which make use of the viscoelastic behaviour of the polymers, are, however, rather challenging.

The production of bicycles in India may approach 150 lakhs per annum before the turn of the century; the production of two wheelers will be about 30 to 40 lakhs during the same period. Engineering plastics can play an important role in improving bicycle designs. The road and load conditions in India are quite peculiar. It is important to recognise that, unlike the western world, these have to be designed for 100 to 200 per cent more load - a bicycle often carries a whole family of four along with some belongings! Thus the mechanical peformance requirements have to be stringent, calling for engineering design modifications and novel materials technology.

Apart from the clear role of superabsorbent polymers in sanitary towels and diapers, the potential impact in agriculture, horticulture, etc., particularly for the arid and semi-arid zones of India is immense. We need highly innovative work in this area. The social impact of such materials, which are difficult to make, can be really great not only in India but perhaps in other Third World countries too.

Polymer additives in concrete can open up a vast area as our cement production will probably reach 80 million tpa by the turn of the century. Since we have a coastline of more than 5000 km, and corosive conditions are generally prevalent, concretes with fibre reinforced polymer materials will have a great scope. Here the use of 'polymer concrete', using unsaturated polyesters, epoxy and polyurethane resins is quite relevant. It is likely that cheap polymers and copolymers based on ethylene-vinyl acetate or styrene-vinyl acetate may be adequate. The problems of adhesion, dispersion and water drainage are quite complicated and need to be studied to guide proper selection of additives.

11   Biotechnology

This area is attracting attention globally. It is an 'equal opportunity' area. We ought to realise that the acquisition of technology in this area from overseas, particularly in pharmaceuticals and human care products, or even agriculture, would be extremely difficult even when compared to high-tech areas in the chemical industry.

The area of separation technology will have to be developed in a major way; otherwise, many breakthroughs may be thwarted in the absence of effective separation processes. Two problems of special interest to India, with its large vegetarian population, may be mentioned.

● *Production of edible proteins to supplement normal sources* ; in the case of microbial proteins, separation of undesirable constituents like nucleic acids by cost effective methods would be required.

● *Production of microbial rennet* , such as the one obtained from the liver of calves, is not acceptable. Similarly, engineered insulin as the usual source of animal origin is not favoured.

The need to develop diagnostic kits for tropical diseases like enteric fevers and malaria is great. Further, the development of enzymes for treating the cow's and the buffalo's milk, where the cattle is fed with silage, requires special consideration.

The large scale exploitation of tissue culture is expected to demand novel sparged reactors, and we should urgently acquire expertise in this area.

The treatment of cellulosic wastes for upgradation, rather than saccharification for eventually producing alcohol, should be a major endeavour. Various woods and agricultural residues generally contain 30 per cent to 50 per cent cellulose, 18 per cent to 35 per cent hemicellulose, and 15 per cent to 25 per cent lignin. It is of great value in India to recover lignin by using innovative strategies.

It is quite likely that less spectacular techniques compared to those based on genetic engineering, such as cell fusion and the immobilisation of enzymes or whole living cells on polymeric substances to give a long lasting biocatalyst, can contribute in a significant way. We need to explore these strategies urgently.

12 Renewable Versus Non-Renewable Raw Materials

This is very popular subject all over the world. India's organic chemical industry was born on the renewable raw material alcohol, which, in turn, is based on sugar cane molasses. It is ironical that before 1956 sugar companies used to pay transport contractors to lift molasses from their factories to avoid a pollution problem. In the recent past, on the other hand, we had no molasses to offer! Thus alcohol as a feed-stock (based on molasses or starches) holds no future in India. Only high value added products such as citric acid, biopolymers based on molasses, etc., will have a bright future. Sugar can be imaginatively used for value added products and there is a good potential for polysaccharides.

There are strong reasons to believe that a substantial part of the consumption of cane sugar will be diverted to high fructose syrup (HFS), as, here, a short duration starch crop can be gainfully employed to provide a value added product. In many locations in India fallow land can be utilised for the production of a high yielding variety of tapioca.

The conversion of cellulosic material to alcohol, after about 75 per cent loss is weight, holds no promise in India for a variety of reasons associated with the capital and recurring costs.

The most obvious outlet for 100 million tonnes per annum of bagasse is for writing paper, and subsequently newspaper. India can possibly emerge as an exporter of writing paper. The involvement of chemical industry in this sector is, however, heavy.

13 Coal Based Chemicals

India's experiments with coal-based fertilizer plants have not been rewarding. Capital costs are very high and the reliability of uninter-rupted operation is not high. No new fertiliser plant is expected to be based on coal upto the year 2001. The use of synthesis gas, based on coal, for the chemical industry is also most unlikely for similar reasons.

However, the prospect of gasification of coal for the power industry, possibly with a combined cycle, is exceptionally bright in view of large scale demand for power and the high ash in Indian coals (approaching values as high as 25 to 30 per cent). Yet another distinct possibility is the exploitation of the very large underground deep deposits for coal - some of them in areas far removed from the coal mines. The *in-situ* gasification of coal, which is a difficult subject, is expected to emerge in a big way, particularly in Mehasana in North Gujarat.

14 Software in the Chemical Industry

In India there is a particular need for increased emphasis on modelling and simulation as a vital tool for process development and scale-up. This is so in view of the fact that, unlike the developed countries, we cannot afford expensive and large-scale pilot facilities. Such a trend is already apparent in the areas of development technologies for reforming, FCC and hydrocracking. We may well be in a position to export know-how, in terms of design and optimisation software.

The export of application software for usage in chemical industries can indeed be a major area of thrust. For instance, software for computer control of precipitation of a variety of materials, like MnO_2, Fe_2O_3, etc., may be very welcome. The simulation of existing plants in India and overseas can be undertaken with some finesse and the potential for domestic use and export is high. The development of methods based on artificial intelligence (the so-called knowledge engineering) to control harzardous plants is a very challenging area, where also there are great opportunities.

15   Major Constraints in the Chemical Industry

It should be emphasised that the chemical industry, and especially petrochemical industry, is capital intensive and it has a long gestation period. The persistent resource crunch has affected the growth of this industry. An adequate resource allocation will, in future, be essential for achieving the planned growth.

Another important constraint, which is likely to persist for some years, perhaps even upto the year 2001, is the shortage of power. The interruptions in power supply and the fluctuations from the standards have been quite detrimental for the industry. A possible solution, partly or wholly, in many chemical plants, is to adopt the total energy concept so that steam requirements and power demands are matched properly. There is great scope for this in industries like fertilizers, soda ash, crakers for olefins, etc. There is scope for improving sulphuric acid plants and, apart from the surplus power generation in these plants, imaginative use of the export steam can be made by proper integration with other plants. We have made some progress in this direction but we ought to do much more.

The growth of the chemical industry could be enhanced considerably if the finished products are produced at competitive prices, and, for achieving this objective, apart from economies of scale, the basic input raw materials and energy costs will have to be brought down to the international level.

16   Concluding Remarks

The Indian chemical industry can catapult itself into one of the leaders in some selected sectors of the chemical industry, notably all types of fertilizers, synthetic fibres and speciality chemicals. R&D in India is expected to make an impact on the world scenario. India is expected to enter the world market in an important way to design, commission and operate chemical plants, and provide service on a decent scale.

17   Acknowledgement

The comments on this report received from the Planning Commission, Department of Chemicals and Petrochemicals, Department of Fertilizers as well as some fertilizer production and engineering consultancy organisations are gratefully appreciated.

13

Water Transport in India

In olden times, water routes were preferred to land
routes, and used to be the main channels of social,
cultural and political interaction. Unfortunately, we
have, in recent times, ignored this available asset.
But it is not yet too late. It is still possible to effectively
exploit water transportation systems through proper
and comprehensive planning.

Contents

13

Water Transport in India

Preamble

In spite of an extended coastline and long navigable rivers, water transport has not flourished in our country. In comparison, the traffic along road, rail and in the air has grown tremendously. What are the socio-economic factors that have prevented the growth of water transport? What benefits will accrue if water transport is consciously developed? To what extent are R&D efforts needed to overcome the difficulties of development? What are their implications for marine ecology and the overall environment?

These were some of the issues discussed by the Expert Group (EG) on water transport in India, constituted by the SAC-PM. The EG held three meetings, in Bombay, Goa, Cochin, to discuss various issues such as

* coastal transport including the inter-island traffic
* inland water transport
* urban water transport and
* development of water sports and tourism, etc.

This document is based on the papers prepared by individual members of the EG. It comes up with specific suggestions and action points which may be found useful by the country's planning authorities. ■

1   Introduction

It has been widely recognised that water transport is a highly energy-efficient mode of transport. In spite of this fact, transport, by both coastal shipping and along inland waterways in India, has been declining over the years, in sharp contrast to the rise of traffic on the railways and road. With a long coastline of 5660 km, and about 14500 km of navigable waterways, India does have considerable potential for an increase in the use of this water resource for transport purposes.

Coastal traffic between ports in India (including the Andaman & Nicobar Islands and Lakshadweep) has been reserved for Indian shipping. But the traffic carried through the 10 major and 165 intermediate minor ports has not grown at all in the last 40 years or so. In fact, but for the increase in movement of off-shore crude oil and coal, the coastal traffic would not have existed. Salt and 'general cargo' traffic fell from a level of 4.71 lakh tonnes and 12.19 lakh tonnes in 1952 to 1.34 and 0.81 lakh tonnes in 1983. Crude oil now accounts for about 137.8 lakh tonnes and finished petroleum products for about 41 lakh tonnes. The coal traffic now carried by coastal ships from Haldia, Paradeep and Vizag to Madras and Tuticorin is essentially for power houses. Table 1 gives the details of coastal shipping traffic.

One of the factors which has been inhibiting the movement of coastal cargo, particularly general cargo, salt and coal, is the high cost of handling at ports. Hence, although the 'line-haul' costs of carriage between ports may be low, as compared to road or rail, when the handling costs are added, the advantage is nullified. It is therefore necessary to find ways to cut down costs of handling coastal cargo.

Traffic carried by inland waterways adds up to about 15 million tonnes of originating tonnage (3 per cent of total) and 1.17 billion tonne kilometres (1 per cent of total transport effort). The iron-ore traffic of Goa accounts for about 96.5 per cent of originating tonnage. About 2 per cent of tonnage gets moved on the Hooghly/Brahmaputra rivers and the balance, largely in Kerala and Maharashtra. In other States, the rivers do not carry much freight traffic.

Table 1
Coastal Shipping Traffic 1951-1987

Freight carried (in lakh tonnes)

Year	Dry Cargo				Wet Cargo	Total Cargo	Passengers carried (in lakhs)
	Coal	Salt	General	Total			
1951	7.7	4.5	12.9	25.1			13.4
1956	11.0	4.8	10.2	26.0	11.8	37.8	9.4
1961	13.7	4.7	15.0	33.4	20.9	54.3	9.6
1966	7.0	3.3	15.0	25.3	30.6	55.9	9.2
1971	5.1	5.2	6.1	16.4	27.3	43.7	5.4
1976	5.7	2.3	1.5	9.5	23.4	32.9	3.9
1977	6.6	1.1	1.3	9.0	23.9	32.9	3.0
1978	6.7	2.2	2.0	10.9	24.2	35.1	3.1
1979	4.7	2.9	1.8	9.4	-	-	3.2
1980	7.7	2.5	1.6	11.8	27.1	38.9	3.8

Table 1 (*Contd.*)

| Year | Freight carried (in lakh tonnes) | | | | | | | Passengers carried (in lakhs) |
| | Dry Cargo | | | | Wet Cargo | Total Cargo | |
	Coal	Salt	General	Total			
1981	12.03	0.88	0.86	13.77	32.01	45.78	3.75
1982	16.71	1.30	1.85	19.86	26.71	46.57	4.36
1983	19.58	1.34	0.81	21.73	33.12	54.85	3.95
1984	17.67	0.64	-	22.01	43.57	65.58	3.81
1985	23.35	-	-	30.35	63.46	93.81	3.81
1986	23.01	-	-	30.01	67.45	97.46	4.25
1987	44.11	-	-	44.11	101.19	145.30	4.17

Source: 1951-78 Dry and wet cargo figures taken from N.T.P.C., P.No. 266

1979-80 Wet cargo figures taken from D.T.R.

1979-80 Dry cargo figures taken from D.D.G. report

1981-87 Dry and wet cargo figures taken from D.G. Shipping's note.

An important factor which inhibits the growth of traffic in both coastal and inland vessels is the *comparatively slower speeds and higher transit times* by water. These lead to high inventory carrying costs which often acts as a factor determining the choice of transport mode. It should, therefore, be possible to channelise bulk commodities, which have an inherent low carrying cost, and which have to be stored for seasonal or perennial use by the slower water transport, by deliberate policy measures which do not add to costs.

To facilitate the carriage of traffic by the Inland Water Transport (IWT), it would be necessary to develop the river infrastructure. The rivers need to be dredged and trained to ensure a minimum draft of water. Facilities by way of terminal/handling arrangements etc. have to be provided. These would require to be developed over short or long-term horizons, and suitable backward and forward linkages, with allied or associated sectors, would have to be considered. For instance, the requirements of irrigation flows in rivers/channels would be factors determining the long-term development of navigational channels. Whilst there may be (and are) cases of excessive water-flow in some of our rivers (excessive in relation to navigational needs), this happy situation is not spread all over the country. A long-term assessment of the potential and requirements for water for hydro-power, irrigation and transport needs to be done.

The IWT Sector's importance in the transport sector was recognised in the Seventh Plan with a provision of a quantum jump in the Plan outlays. Table 2 gives the position regarding public sector expenditure in this sector.

Table 2

Successive Plan Outlays for IWT (Rs in crores)

Plan-period	III	IV	V	VI	VII
Amount	4	11	16	70	226

The Inland Waterways Authority of India (IWAI) was constituted in the Plan period to foster the development of this hitherto neglected sector. A portion of the river Ganga has been declared as a national waterway, and some other rivers have also been identified for the purpose. This improvement in the institutional set up should accelerate the development of the waterways and the water transport fleet on sound lines.

The development of waterways essentially means the removal of navigational hazards like narrow width and shallow water, prevention of siltation and bank erosion. Adequate vertical clearances need to be ensured at bridge crossings. Night navigational facilities are also called for.

In addition to long distance water transport, there is considerable scope for commuter traffic in rivers, bays and estuaries within larger cities like Bombay, Calcutta, Goa, Cochin, etc. The problems of traffic congestion in urban areas can be simplified by exploring the water alternative wherever possible.

As to river-craft, there is a need for appropriate technological upgradation. Vessels which can adapt to the availability of wind energy, and float in a minimum of draft, are necessary. Mechanisation of craft, consistent with the expertise available for maintenance, also needs to be fostered. A greater awareness of fuel-efficiency, and the prevention of environmental pollution of water by fuel emissions, needs to be inculcated, whilst applying borrowed technology. There is also the need for standardisation of shore-handling equipment to facilitate carriage of larger sized parcels, so that the unit handling costs at interface points can be reduced. In the field of river passenger craft, there is a need to develop fast boats along riverable travel routes and ferries. Standards of safety and comfort would also need to be reviewed afresh.

In the case of coastal shipping, technological inputs to reduce the handling time/costs at terminals would appear necessary. This would require a re-alignment of facilities on ships, and on-shore, in order to achieve at a higher parcel load per handling operation. Containerisation of package to get at a load of, say, 5 tonnes appears to be one way. The linkages at intermediate and minor ports with road/rail and IWT modes would also need to be reviewed and improved.

Water transport in certain tourist centres can provide added attractions through cruises, water sports, amusement areas etc. At a time when the need for encouraging tourist trade is being felt strongly, the demonstrated contributions abroad of water transport should serve as useful guides to us here.

In all these issues one thing stands out clearly: there are tremendous benefits possible from a judicious expansion of water transport provided certain basic problems are solved, new amenities are created and the overall infrastructure improved. This is where R&D efforts are called for to make the best use of science and technology.

This chapter highlights the role of such efforts. In the succeeding sections, the issues referred to above are elaborated in detail for focussing attention on (a) the benefits that would accrue from the expansion of water transport and (b) the problems that have to be overcome to achieve the required goals. The S & T inputs are outlined next. Finally, the report also touches on certain strategic aspects of water transport.

We briefly discussed strategic aspects of water transport also. Two specific issues, according to us, deserve closer scrutiny. These are outlined in Appendix-1.

2   Coastal Shipping in India

2.1   Introduction

India has a vast coastline of about 5,660 Km and is bounded on three sides by the Arabian Sea, the Indian Ocean and the Bay of Bengal. The country is bestowed with numerous natural ports, and the Central and the State Governments have also set up several ports at strategic points to enhance the sea-borne trade. Therefore, coastal shipping in India can be developed in India with very little effort. Such a development is also quite essential because there are a number of islands, on both the east and west coasts, whose only link with the mainland is via sea routes.

During the past few decades, fortunately, coastal shipping has started playing an important role in meeting the country's domestic transportation needs. Coastal shipping has also come to be regarded as an adjunct to the country's international shipping. Coastal shipping may be divided under three heads:

● mainland coastal shipping.
● mainland-island shipping.
● interisland shipping.

2.2   Growth and Development of Indian Coastal Trade

Coastal movement in the fifties, sixties and seventies comprised mainly general cargo, coal, salt, and crude oil and petroleum products. The general cargo in turn comprised food grains, jute, tea, iron and steel products, fertilizer, cement etc. However, in the eighties the coastal trade has generally been limited to coal, crude oil and petroleum products and, to a certain extent, to clinkers.

In the fifties and sixties large quantity of general cargo was being transported on the Indian coast. However, this trend declined steadily from about 15 lakhs tonnes in the mid-sixties to about 1 lakh tonnes in early eighties. From the mid-eighties there is practically no movement of general cargo by coastal ships. The reasons for the decline are:

● roadways and railways provide a quicker mode of transport, and quite often, (in the case of roads) even a door-to-door service. Therefore as these alternative modes of transport developed, traffic from coastal shipping was diverted to them.

● due to high cargo handling cost at the ports, the coastal shipping became uncompetitive.

● due to the delays involved in clearing of goods at the portheads, and due to increasing documentation formalities at the ports, coastal shipping was unable to attract high freighted cargo which moves in small parcels.

During the fifties and the mid-sixties, there was large movement of coal, largely for the Railways, along the Indian coast. However, from the mid-sixties to the late seventies the coal movement dropped down drastically from 16-17 lakhs tonnes to 2-4 lakhs tonnes due to following reasons:

- introduction of diesel locomotives by the Indian Railways
- electrification by the railways
- increase in coal transportation by the rail mode.

From the early eighties the movement of coal along the Indian coast started growing rapidly and, by 1987, reached a level of almost 48 lakhs tonnes per annum. The increase is because of installation of coastal thermal power stations in South India.

At present thermal coal is basically shipped from Haldia port. A limited quantity of coal is also being shipped from the Visakhapatnam and Paradeep ports. Shipping of coal from the latter two ports is, however, limited because of the non-availability of automatic coal loaders at these ports.

New thermal power stations are proposed to be set up at various other coastal cities of India such as Visakhapatnam, Pondichery etc. and therefore there are indications that, by the turn of the century, movements of coal along the coast would go up multifold.

Several studies have been carried out in the past to determine the relative economics of transportation of coal by coastal ships versus rail routes. Generally these studies have shown that it is economical to transport coal by coastal shipping, especially when the destination of the cargo is near the coastline.

So far as salt is concerned, till the beginning of the eighties, a large quantum of it was moving along the Indian coast. This was basically moving on coastal ships as a back-haul cargo, on collier ships as well as (because of insufficient capacity) on the rail route. The Central Government was subsidising the freight movement of salt by coastal ships. However, after the withdrawal of the Government subsidy in the early eighties, the movement of salt by the coastal ships declined sharply and, at present, only about 2,000 tonnes of salt is moved by coastal ships. It is estimated, however, that during the Eighth Plan period, a total of 0.30 lakhs tonnes of salt is likely to be moved by the coastal ships.

Table 3 indicates separately the movement of total cargo (general cargo, coal and salt) on the west and east coasts of India. It may be noted that the survival of coastal shipping on the east coast is mainly due to movement of thermal coal.

Table 3

Total Cargo (General Cargo, Coal and Salt) Movement Coastwise

Year	West Coast of India (lakh tonnes)	East Coast of India (lakh tonnes)	Total (lakh tonnes)
1952	7.00	19.26	26.26
1953	6.95	21.19	28.14
1954	6.33	22.61	28.94
1955	6.83	20.20	27.03
1956	7.10	18.74	25.84
1957	7.26	18.41	25.67
1958	6.42	20.10	26.52
1959	6.91	18.42	25.33
1960	6.26	20.78	27.04
1961	7.53	24.69	32.22
1962	8.08	27.31	25.39
1963	7.74	29.79	37.53
1964	7.17	24.37	31.53
1965	7.85	20.33	28.18
1966	8.65	11.23	19.89
1967	8.74	10.71	19.45
1968	8.90	8.15	17.04
1969	7.19	8.19	15.39
1970	5.28	4.03	9.31
1971	5.69	7.86	13.55

Table 3 (*Contd.*)

Year	West Coast of India (lakh tonnes)	East Coast of India (lakh tonnes)	Total (lakh tonnes)
1972	7.20	9.72	16.92
1973	5.02	8.09	13.11
1974	4.52	7.47	11.99
1975	6.69	10.42	17.12
1976	2.09	7.59	9.69
1977	2.09	5.49	7.58
1978	2.94	9.48	12.42
1979	1.84	7.58	9.43
1980	1.41	10.39	11.81
1981	0.89	12.89	13.77
1982	0.98	18.87	19.86
1983	0.96	20.78	21.73
1984	0.12	18.19	18.31
1985	0.07	23.35	23.42
1986	0.05	38.00	38.05
1987	0.02	48.00	48.02

Source: From Rajwar Committee Report dated September 1987.

The movement of cement clinker on coastal shipping is basically limited to the west coast of India. Though a small quantity of this commodity is moving on coastal shipping, this could be considered as a potential cargo source for coastal ships.

At present India's offshore production of crude oil is limited to Bombay High fields. The total quantity of crude oil produced in the Bombay High offshore fields is now being utilised by the Indian refineries. Out of the total quantity, some part is transported through pipelines and, the other part is transported by the coastal ships. Table 4 indicates the quantity of crude oil moved by coastal ships. It is expected that, during the Eighth Plan period, a total of about 140 lakh tonnes per annum will be transported to the Indian refineries by the coastal ships.

Table 4

Coastal Movement of Crude Oil and Petroleum Products

| Year | Quantity Carried (in lakh tonnes) | |
	Crude Oil	Petroleum Products
1982/83	49.34	30.61
1983/84	67.53	33.12
1984/85	82.39	41.05
1985/86	124.24	45.00
1986/87	137.84	50.00
1987/88	135.00	58.00

Source: Rajwar Committee Report of September 1987.

It may be noted that, during the initial years, the crude oil produced at the Bombay High fields was exported out of the country, but, with the installation of more refineries in the country, transportation of crude oil to Indian refineries has increased, which in turn has increased the coastal shipping. This, however, has the opposite effect on the movement of petroleum products. The total coastal movement of petroleum products is at present about 60 lakh tonnes. However, in the coming years, the movement of petroleum products along the coast is expected to drop down to a level of about 25 to 27 lakh tonnes due to the commissioning of more oil refineries on the Indian coast.

2.3   The Movement of Passengers

Though India has a large coastline, there is practically no passenger traffic along the coast. The last two passenger ships which operated between the ports of Bombay and Goa, and some intermediate ports, have been withdrawn from service since August 1988 following a decision taken by the Central Government. It was observed that the passenger service on this Konkan coast route was incurring heavy losses and it was becoming difficult for both the Central and the State Governments to bear such losses.

Since a pure passenger service was found to be unremunerative, consideration is being given to commence a Ro-Ro freight service on the west coast of India between Jafrabad port in Saurashtra and New Mangalore port in Karnataka. The vessels would call at intermediate ports like Bombay, Ratnagiri, Devgad, Marmagoa etc. The carriage of a suitable number of cargo trucks on the Ro-Ro ferry service is expected to compensate for the losses incurred by a dedicated passenger vessel.

The commencement of high speed passenger service, by means of fast speed passenger ferries, on the Bombay-Kutch and Bombay-Goa routes is also under consideration.

2.4   Growth and Development of Mainland-Island Trade

There are two main groups of islands of India: the Andaman and Nicobar (A & N) group of islands (on the south-east) and the Lakshadweep group of islands (on the south-west).

Shipping is the lifeline for these islands. Though there are some air services between the mainland and these islands, ships are considered to be the basic link. Therefore, a high proportions of passengers and cargo is transported on sea routes to and from both these group of islands. It is well-known that the passenger and cargo transport demand in any region is strongly correlated with the population and the socio-economic characteristics of the region.

The passenger traffic between the mainland and the A & N islands, as well as the mainland and the Lakshadweep islands, by sea routes, has been increasing from year to year.

Presently there are four passenger vessels (Indian ownership or chartered) operating on the A & N sector and two passenger vessels (Indian ownership) operating on the Lakshadweep sector. As some of the ships on the A & N sector have to be scrapped in the near future, orders for three new ships have been placed.

The cargo on the A & N islands sector is moved by cargo ships and cargo-cum-passenger ships. However, on the Lakshadweep sector, the cargo from the mainland is moved by cargo vessels as well as mechanised sailing vessels and the cargo from the islands is generally moved by the mechanised sailing vessels. Since some of the vessels on the A & N sector have to be replaced in the near-future, orders for new ships have been placed. Similarly on the Lakshadweep sector orders for four cargo vessels are being placed to replace the old ones. Special attention has been given to the draft of these vessels so that the cargo vessels could go right to the island jetties for discharge.

2.5   Inter-Island Services in the A & N Sector

The inter-island services in the Andaman & Nicobar group of islands can be sub-divided under two heads:

● Inter-island service
● Foreshore service.

The inter-island services include the routes where the ships have to travel from one island to the other islands extending all the way from the North to the South. On the other hand, the foreshore services include only those routes where the ships operate in sheltered waters.

Presently there are five passenger-cum-cargo vessels on the inter-island passenger and cargo service in the A & N sector. Orders for nine new vessels with different passenger and cargo capacities have been placed on various Indian public sector ship-yards. These are expected to be inducted during the next few months on the inter-island passenger and cargo services of the A & N sector.

There are four passenger-cum-cargo vessels, at present, on the foreshore passenger and cargo service in the A & N sector. Orders for 12 new vessels, with different passenger and cargo capacities, have been placed on various Indian public sector shipyards. These are expected to fructify in the course of the next few months for induction on the foreshore passenger and cargo services of the A & N sector.

There are no dedicated inter-island passenger vessels in the Lakshadweep group of islands. The passenger vessels to the mainland also serve the inter-island passenger traffic. It has been decided that, in future, the two mainland-island passenger ships would call at only two or three main islands and, thereafter, the passengers shall be transported to the other islands by inter-island passenger vessels, to be acquired shortly.

To cater to the inter-island passenger traffic, orders for two fast speed passenger vessels are being placed, and the vessels are expected to be inducted into service by early 1990.

2.6   Projected Scenario in the 21st Century

2.6a   Mainland Coastal Shipping

Looking at the coastal shipping trend in the last several years, it appears that the main cargo which could be expected to be moved on sea routes by the year 2000 AD would be limited to coal, cement clinker, crude oil and petroleum products. It is quite likely that there would be a certain movement of passengers on coastal routes, subject to special considerations given by the Central and State Governments, and by the shipping companies. The shipping traffic of general cargo in the break-bulk form is not expected to survive in a pattern similar to the past or the present, for reasons mentioned earlier. Apart from this cargo, there would be a large movement of natural gas, and possibly of iron ore concentrate, by coastal ships from Mangalore to the Vizag steel plant.

It is also expected that the import and export of general cargo would be containerised to a large extent. To support this, the transhipment of containers to and from the other smaller ports would automatically be required.

2.6b   Mainland-Island Shipping

The passenger traffic on the mainland A & N sector is expected to increase steadily both on account of increase in population as well as increase in tourism. Though a large fraction of the traffic would be attracted by the airlines, it is expected that, by the turn of century, there

621

would also be an annual passenger traffic of about 2.4 lakhs by the sea routes.

The cargo transportation by sea route would also increase steadily in a way similar to the passenger traffic. However, the movement of timber traffic, from the islands to the mainland, would either stagnate at a level of about 20,000 cubic metres per year or even decline further.

In the Lakshadweep sector, the passenger traffic from the mainland is increasing at a very fast rate and, by the turn of the century, passenger traffic by sea is expected to go up to as much as 1 lakh per annum.

As regards transportation of cargo, to maintain realistic sailing schedules, a decision has been taken to bifurcate the passenger traffic from the cargo traffic. So far the cargo was being discharged outside the lagoons of the individual islands. This used to result in heavy loss of cargo. A decision was therefore taken to acquire low draft cargo vessels so that the cargo transportation can be done independent of passenger transportation. Also, that the cargo ships could then go inside the lagoons and discharge the cargo at the jetties of the individual islands.

2.6c   Inter-Island Shipping

It has been observed that both passenger and cargo movement have been steadily increasing both on the inter-island routes and the foreshore routes of the A & N group of islands. To take care of this steady growth in the traffic, orders for additional cargo and passenger ships have been placed. Further vessels may also be required in the near future.

The inter-island passenger traffic in the Lakshadweep group of islands has increased during the last few years and, because of the decision taken to terminate the mainland-island passenger ship at 2/3 main islands only, the inter-island passenger traffic in Lakshadweep is expected to increase at a very fast rate in the years to come. Orders for new fast speed inter-island passenger vessels have been placed.

2.7   Recommendations

In view of the above mentioned future needs, and keeping in mind the performance of coastal shipping both in the past and the present, the following recommendations are made:

● Efforts should be made to transport the maximum quantity of thermal coal on coastal sea routes.

● To match the international trend of moving the cargo by container ships, our country should also take fast steps towards the container trade and establish an efficient container feeder service on coastal sea routes.

● Efforts should be made to commence passenger services through coastal shipping.

● Natural gas being one of the most important sources of energy, and in view of the fact that this is available in abundance in the offshore fields of the country, a full investigation should be made to explore means of transporting it along along the coastal sea routes in a liquified form.

● Efforts should be made to transport crude oil and petroleum products along the coast in the most efficient manner.

● The A & N and the Lakshadweep group of islands being an integral part of our country, all possible effort should be made for efficient transportation of men and material to and from the mainland-islands as well as on the inter-island sectors.

● The Central Government must lay more stress on a higher participation by the public sector shipping companies in coastal shipping.

● The R&D effort, needed to meet the requirements listed above has to be located within dedicated groups in universities and national laboratories, and with close interaction with shipyards, shipping companies and port authorities.

3   The Inland Water Transport System

3.1   Introduction

India has a number of big river systems like the Ganga, the Brahmaputra, the Mahanadi, the Godavari, the Krishna, the Narmada and the Tapti. However, although a length of 5200 km of major rivers of India is navigable by mechanised crafts, the length actually utilised at present is only about 1700 km. As regards canals, India has a length of 4300 km of navigation canals out of which only a length of 485 kms. is suitable for mechanised crafts.

No accurate estimate of the traffic carried by water transport is available. However, an estimate of the cargo carried by mechanised vessels on some important waterways, and by some prominent operators, is made in Table 5.

Inland water transport (IWT) is recognised as the most suitable mode of transport for transporting bulk and semi-finished commodities because of its very low consumption of energy, low relative cost of operation, very low pollution effect and its negligible needs of useful land. However, for various reasons, its growth in India has been poor so far.

Since Independence, the Central Government has been considering the question of restoring IWT to its rightful place in the transport system of the country. The problems relating to inland water transport were examined in the past by the Estimates Committee of Parliament (1956-57), the Gokhale Committee on Inland Water Transport (1959), the Committee on Transport Policy Coordination (1966), the Estimates Committee of Parliament (1970), the Estimates Committee of Parliament (1974-75) and National Transport Policy Committee (1980). In pursuance of the recommendations of these Committees, some progress has been made in regard to the development of this mode of transport, and the Planning Commission had made increasing allocations for the IWT schemes. The provisions made, and the actual expenditures, incurred on the schemes upto the end of the Sixth Five-Year Plan are given in Table 6 (see pages 626-627).

Table 5

Existing Distribution of Traffic by Waterways and States

Waterways system	State	Volume of traffic (million tonnes)	Average Lead Kms.	Traffic in Million tonne Kms	% age of total IWT traffic (tonnes)	% age of total traffic (tonne Kms)
Mandovi-Zuari	Goa	14.76	50.0	738.0	96.5	75.5
Hooghly Brahma-putra River	West Bengal and Assam	0.31	750.0	230.0	2.0	23.5
Rivers & Creeks of Maharashtra	Maharashtra	0.07	27.0	2.00	0.5	0.2
West Coast Canal and Vackkaiers of Kerala	Kerala	0.15	56.0	8.00	1.0	0.8
		15.29		978.0	100.0	100.0

Source: From the report ''All India Transport Study (Inland Water Transport)'' prepared by CES for the Planning Commission during the year 1987.

625

Table 6

The Funding Status of IWT in the First Six Five Year Plans**

Five Year Plan	Central Schemes	Centrally Sponsored Schemes	Total
First Five Year Plan			
Provision	Nil	Nil	Nil
Actual Expenditure (for GBWT Board)	33.08		33.08
Second Five Year Plan			
Provision	143.32	Nil	143.32
Actual Expenditure	72.34	Nil	72.34
Third Five Year Plan			
Provision	438.18	322.00	760.18*
Actual Expenditure	126.62	125.71	252.33
Annual Plan 1966-67			
Provision	164.52	32.62	197.14
Actual Expenditure	54.09	27.62	81.71
Annual Plan 1967-68			
Provision	192.40	33.65	226.05
Actual Expenditure	299.07	27.98	327.05
Annual Plan 1968-69			
Provision	99.00	31.00	130.00
Annual Expenditure			130.00

Table 6 (*Contd.*)

Five Year Plan	Central Schemes	Centrally Sponsored Schemes	Total
Fourth Five-Year Plan (1969-74)			
Provision	500.00	400.00	900.00
Actual Expenditure	369.89	303.91	673.80
Fifth Plans 1974-78			
Provision	1392.18	1100.00	2492.18
Actual Expenditure	539.82	446.86	986.68
Annual Plans 1978-79 & 1979-80			
Provision	2300.00	300.00	2600.00
Actual Expenditure	176.89	86.25	263.11
Sixth Plan 1980-85***			
Provision	4117.00	383.00	4500.00
Expenditure	3821.00	84.00	3905.00

*   Expenditure was to be restricted to Rs. 400.00 lakhs.
**   From the report of the Working Group on IWT for the VIIth
     Five Year Plan.
*** From draft report of the Working Group on IWT in VIII Plan.

The Central Government has, thus, been showing a sustained interest by continuing the enhancement of plan funds to encourage IWT as a supplementary mode of transport. Various committees have met from time to time with recommendations for action plans. Yet much remains to be done, especially at the level of translating recommendations into action. The record of the various State Governments in this respect is much less satisfactory, as shown, for example, by the lack of utilisation of funds earmarked for the various IWT programmes, even in the centrally sponsored schemes.

Of course, to establish the feasibility and viability of IWT, pilot studies are necessary. To illustrate the kind of results that such studies can come up with, the next section summarises the studies on IWT conducted by the National Transportation Planning and Research Centre (NATPAC) at Trivandrum.

3.2   Studies on IWT Conducted by NATPAC

Study on 'Employment Potential of Inland Water Transport in Kerala', sponsored by the Planning Commission, Government of India

This study assessed the employment potential of the inland water transport system (which hitherto was not available) and the results were subsequently used by the National Transport Policy Committee of Planning Commission while formulating the National Transport Policy. The study revealed that country crafts used for passenger transport provided employment to the tune of 58.8 man years for an investment of Rs 1 lakh. The employment, per Rs 1 lakh of investment, worked out to be 20 man years in the case of non-mechanised freight transport, and about 16 man years for the total inland water freight traffic in the study area of the Cochin region. The study showed that the construction of canals yields quite a large employment (over 13 man years for investment of Rs 1 lakh). The indirect employment generation potential, such as boat building and craft making industries, has been estimated at 14.3 man years per Rs 1 lakh investment.

The total employment generated in IWT is estimated at 33.59 man years, per lakh of rupees investment, as compared to 16.93 in the case of road transport and 4.3 in the case of railways.

'Intermodal split of traffic - A case study of Cochin-Quilon region', sponsored jointly by the Department of Science, Technology and Environment, Government of India and the Government of Kerala

This study was conducted to assess the quantum, and the reasons thereof, for the choice of mode for passenger and goods transportation in a region where road, rail and water transport systems operate side by side. The total goods traffic in the region was estimated to be 18.13 million tonnes in 1979 and the share of IWT was 1.55 million tonnes. The share of IWT in the passenger transport was also very marginal (1.67 per cent). The study also showed that, in the given socio-economic conditions of the people in the region, and in the case where road, rail and IWT system are in co-existence and are perfect substitues, roadways will be preferred for making 111548 passenger trips, railways for 48477 trips and water transport for 3450 trips. The reason for choice in each case varies from better service reliability, travel time and travel cost. The conclusions of the study were used for planning the transport system in the state for various time periods.

Study on pattern of goods traffic by road and IWT in Kerala, sponsored by the High Level Committee on Physical Infrastructure and Transport appointed by the Government of Kerala

The study was conducted to assess the pattern of goods traffic, both by road and inland water transport, and also to assess the comparative cost to the economy for moving one tonne km of cargo by road and IWT as compared with rail. The total volume of road based goods traffic in the state was estimated to be 29.21 million tonnes, consisting of 12.59 million tonnes of inter-state traffic and 16.62 million tonnes of intra-state traffic. The total transportation taking place in the waterways of the state by country crafts and by the public sector undertaking (SWTD and KINCO) was estimated to be 2.217 million tonnes involving 73.444 million tonnes km during the year 1983. The comparative cost analysis revealed that the cost, to the economy, for moving a tonne of rice, stone and fertilizer, was less by road for distances upto 100 km, for sugar and cement upto 150 km, for coconut, tea and timber upto 200 km and for rubber and rubber products upto 750 km.

Techno-economic studies of the Cochin-Quilon section of the West Coast Canal in Kerala, sponsored by Inland Waterways Authority of India and the Government of Kerala

The study was taken up as a prelude to declaring the West Coast Canal as a national waterway, and to find out the financial and economic viability of various improvement proposals. The traffic studies, carried out to assess the potentials of IWT, showed that IWT can attract traffic to the tune of 1.71 million tonnes, and that waterways cost less to the economy for transporting a majority of commodities (although road transport is better for collection and distribution of traffic over short distance). Commodities like iron and steel, minerals, lime shells, clay, bricks and tiles, construction materials, electric goods, cashew nut, spices, chemicals and fertilizers, POL etc. have been identified to be transported in this sector.

Based on traffic intensity, 14 terminals have been identified for improvement, and to berth modern vessels, at a cost of Rs 19.57 crores. The dredging cost worked out to be Rs 14.27 crores and the cost of bank protection was estimated at Rs 5.18 crores; Rs 3.82 crores was provided for installation of handling cranes. The financial analysis of the proposed investment of Rs 42.84 crores was made and it was found that the imposition of a levy on goods, in lieu of the capital provisions, collection of wharf and godown rents and other revenues will yield an internal rate of return of about 11.81 per cent. The economic analysis has also revealed that the project is economically viable with an internal rate of return of above 16 per cent.

3.3   Recommendations

3.3a   Organisational and Developmental Aspects

Studies like those described in the previous section demonstrate that IWT has an important role to play in the overall development of the country and that, normally, one would expect the IWAI to play a leading role in this field. However, it has been noticed that the IWAI has, for a variety of reasons, not so far been very effective. We must appreciate fact that IWT is a new area with a lot of exciting possibilities. It is therefore essential that the IWAI receives the special treatment normally given to a nascent industry.

Keeping these points in view the following recommendations are made under three headings: (a) Central Government (b) State Governments and (c) Private Sector. These recommendations relate to specific IWT projects, as well as to the policy guidelines and the infrastructure that are needed to achieve success. Issues relating to urban transportation will be discussed separately in the next section.

Central Government

● Development of national waterways, like Ganga-Bhagirathi-Hooghly, Brahmaputra-Barak, Sunderbans and the West Coast Canal, should be taken up in the immediate phase (1990-95) followed by the development of rivers like Mandovi, Zuvari and Cambarjua Canal, Mahanadi-Brahmani, Godavari-Krishna and Narmada in the second phase (1995-2000)

● Priority should be given for the development of an assured navigation channel with safety measures like navigation aids, including night navigation facilities and the provision of modern cargo handling terminals

● Encourage private participation by extending suitable financial incentives for transportation of cargo, building repair bases at suitable places along the national waterways and the procurement of specialised flotilla for the development of tourism

● Encourage the integrated development of major rivers to cover irrigation, hydel power, flood protection, environmental protection, inland water transport, fishery and forestry etc. by creating river basin management bodies, represented by related interests of development, and also through foreign assistance from countries having previous experience in such developmental activities

● Ensure that IWT is considered as one of the modes of transport for the movement of bulk commodities by a conscious and deliberate policy which will divert traffic along feasible IWT corridors. In this connection, Brahmani, Godavari and Narmada rivers should be

631

developed as the ultimate IWT corridors for movement of coal

● Integrate IWT with coastal movements of bulk cargo. Also,instruct various major ports to keep IWT in mind while preparing their master plans for transiting cargo. Major ports should also allot the berthing front and land area for cargo handling in the ports free of cost, and thereby encourage inland and coastal transport of cargo

● Consider a 100 per cent loan assistance to the States for taking up techno-economic studies and for the improvement of navigability of waterways. Rivers like Sone, Gandak, Ghagra, Kosi, Yamuna and the rivers of the North-east like Barak, Desang and Burhi Dihing are recommended for consideration in the immediate phase 1990-95

● Encourage IWT to be one of the essential factors in deciding location of new industries and ware-houses

● Grant Socio-economic factors the over riding priority, over mere financial returns, while deciding the viability of a scheme

● View the criteria fixed by the National Transport Policy Committee, (1980) for the declaration of national waterways, with flexibility to equally weigh the technical, physical and socio-economic factors peculiar to the particular places through which the waterways pass

● As the Inland Waterways Authority of India has to play a crucial role, not only in the development of IWT infrastructure in the country, but also to function as a research, design and advisory agency to the Centre and the States at the apex level, they may be permitted to decide and create their own well-balanced organisation, with trained manpower, commensurate with the budgetary provisions

● Navigation does not consume water; it only requires an assured availability of water. This is of paramount significance in the lean season, and in the upper reaches of rivers, to keep the channel navigable round the year. There should therefore be a national policy that a minimum quantity of water must be made available (by release of water from barrages/locks etc.) in the upper reaches of rivers to maintain a minimum draught in the navigational channel. This policy will also facilitate the natural process of pollution control and keep the river alive

● For a better and more balanced management of water, particularly in the high flood season, a portion of money from flood control measures could be diverted to conservancy measures, to maintain an assured channel with sufficient depth; this will also greatly help in easing the devastating effect of floods from year to year.

State Governments

● Ensure a serious involvement of the States in the construction, management and maintenance of IWT schemes. The State Governments may be asked to create independent IWT Departments for this purpose; the funds to this may be made available by the Centre as 100 per cent grants

● Create new transport corporations to operate the national waterways, with financial assistance from the Centre in the form of capital subsidy and equity participation. They may also be encouraged to form cooperative societies for channelising central grants and loans for modernising and mechanising the country crafts.

Private Sector

● Acquire and operate vessels in the waterways and also develop their own terminals whenever necessary on the national waterways

● Mechanise the country crafts

● Set up new repair facilities and modernise the existing ones along the waterways.

3.3b   R&D Structures

To emerge as one of the stable modes of transport, the IWT requires inputs from modern science and technology. These areas include river

633

engineering, design of shallow draft vessels, special crafts like coastal-cum-inland vessels, dredging, hydrographic surveys, communication equipment and navigation aids, river training, river cruisers, design of terminals and cargo handling, types of transshipment terminals, and above all, the impact of IWT operations on ecology and pollution control. By balanced inputs of research and design, it would be possible to expedite the pace of development in this nascent area. No head-way can, however, be made in this direction unless the organisational strength is built up in adequate measure, and on sound lines.

We will again identify the R&D structures needed for this purpose.

Central Government

● Create a separate R&D cell, headed by an experienced officer, to scrutinise the R&D proposals, monitor their progress and ensure the implementation of the results of the study. The actual R&D work could be done through sponsoring organisations like CWPRS, Indian Institutes of Technology, CBIP, Shipbuilding Yards etc. As it is essential to progressively develop deep water transport in IWT sector by a systematic river training and control, CWPRS, the premier research station in this sector in the country, may be encouraged to develop physical and mathematical models on the basis of a permanent tie up with the IWAI

● Develop a national inland navigation institute to train the manpower at managerial and senior supervisory levels

● Develop and strengthen the study department to collect data on river parameters, cargo and vessels, and build up a strong database, to enable the planning, design and development of waterways on scientific lines

● Develop real time hydrographic survey capability by automatic data logging; develop computer-based data management for updating the hydrographic data, thereby monitoring the morphological behaviour of the river regime.

State Governments

● Develop regional training institutes to train the crew and navigators with the financial assistance from the Centre

● Collect data on the behaviour of waterways, cargo movement and types of vessels, and maintain a database through exclusive IWT Departments, to be created under 100% grants from Centre. The IWT Departments shall also maintain a close liaison with the activities of flood protection and drought relief works to obtain optimum benefits in an integrated approach.

Private Sector

● Set aside a specific part of their R&D budget for supporting projects relating to IWT. This should be made mandatory on industries that benefit from IWT, directly or indirectly.

4 Water Transportation in Urban Areas

4.1 Introduction

Most of the urban agglomerations in India have their origins rooted deep in history. In ancient days, water bodies and running water courses were considered to be an essential prerequisite for town planning activities. Indeed, this was a common planning parameter for towns and cities all over the world. Water courses were essential not merely for drinking and irrigation purposes, but also preferred as transportation corridors. In fact, water transportation is perhaps the oldest transportation facility known to mankind. During the days when engineering and planning were not fully understood the gift of nature, in the form of running water courses and other water bodies, was gratefully accepted as the corridor of mobility.

It is known how, in olden times, water routes were preferred to land routes, and how these were used as the main channels of social, cultural and political interaction amongst different societies. Civilisations grew along river banks, often from small hamlets. Shahjehanabad was built on the banks of Yamuna, the small villages of Kalighat, Sutanati and Govindpur grew into Calcutta with the river Hooghly as the major spine, the sleepy fishing village on the east coast took the shape of the Madras we know today and, in the west coast, Bombay grew up on an island. European transportation technology has always laid a great degree of emphasis on water transportation and the same was reflected to a large extent in India.

While intra urban travel used to be performed largely along rivers and canals, river front areas were widely used for recreational purposes and as tourist spots. In Kerala the backwaters have been traditionally utilised for transportation, and for water sports. Excellent potential exist for the utilisation of rivers and canal systems in Goa and the north eastern towns, for transportation of men and material, and for promotion of tourism. Any map that shows the navigable stretches of the river system in India, is indicative of the fact that practically all important towns of our country are endowed with river fronts.

However, a lack of far-sighted planning has rendered many of these river stretches unsuitable for navigation. Inadequate maintenance has even robbed the water courses and water bodies of their potentials as recreational areas. The discharge of chemical and other effluents, and the unchecked growth of weeds has affected the quality of water bodies. Deficient maintenance has resulted in shifting of water courses. Bridges across such water courses have brought about head room restrictions in many instances. There has been practically no modernisation in the supporting industry for the maintenance and modernisation of crafts.

In other words, an available asset, to serve as an essential means of transportation, has, due to negligence, now come to be identified as a liability to the society. However, it is perhaps still not too late to restore to the intra-urban water transportation system a part of its past glory through proper and comprehensive planning techniques.

4.2   Urban Transportation Scenario

Some emphasis has been laid, during recent times, on evolving transportation plans for some of the more important urban centres in India. The efforts have been mostly towards working out a comprehensive transportation plan depending primarily on road and rail based systems. The capacities of transportation corridors in most of the metropolitan cities have been totally exhausted and the movement of passengers and goods along such corridors have become increasingly vexing. The Yamuna did not find much importance as a possible transportation corridor in the comprehensive transportation plan for Delhi (1969-72). No comprehensive action plan has yet been drawn up to use the backwaters of Cochin as a part of the total transportation grid. The potential of Mandovi and Zuari still remain to be exploited, at a place where this could provide an excellent alternative to the bus transportation system operated under the Kadamba Transport Corporation. The Buckingham Canal system of Madras has remained in suspension for very long. The recent comprehensive transportation planning studies for Bombay did not really present an action plan for augmenting the water transport system (even though there has been some talk of an elevated west coast expressway)

Movement of goods has usually troubled the transportation network of each city. Wholesale markets have been planned without taking into cognisance the role that could be legitimately played by water courses. Calcutta still continues to have certain types of conventional system of water transportation along the Hooghly. There has been practically no modernisation of this system in spite of the fact that population densities along the banks has registered higher rates of growth. This story repeats itself in practically every other town and city endowed with natural water courses in our country.

Most of the cities suffer from acute shortage of 'lung spaces', which are supposed to be made available through the provision of open spaces, greens and river front areas (water bodies). Hardly any effort has so far been made towards translating planning recommendations, regarding river training activities and river front development, in most of these cities. It is perhaps time to realise that such development measures could become very handy in maintaining a healthy urban environment, and in promoting recreational facilities and water sports, besides generating certain revenue. The salutory effect of deweeding the Dal Lake in Srinagar could be cited as a case in point.

4.3 Present Status of Water Transportation System in Urban Areas

In spite of its inherent potential, water transportation does not occupy a significant position in the system of total transportation within urban areas today. Indeed, barring a few cities like Calcutta on the east coast and Cochin on the west coast, the system of water transportation as a part of urban transportation is practically non-existent in other cities. Even here, only ferry services are run with the help of conventional crafts. In places like Cochin the services are more informal in nature. Even in Goa intra-urban travel through water transport is not very significant. However, a certain quantum of inter-urban movement of goods traffic does take place along water courses in all such cities. The west coast in Bombay does not offer very conducive natural conditions for operating conventional ferry services. However, the sheltered cross channels and the eastern waters in Bombay display a lot of promise for being exploited as transportation corridors.

In Calcutta, passenger ferry services across the river separating Calcutta from Howrah had a natural uptake as congestion went on increasing along the Howrah Bridge. This is the only city which reports time series data at a fairly disaggregate level for passenger ferry services. Some data on goods movement along water courses in the formal sector are also available where such systems are in operation (e.g. Goa). Time series data on movement of goods along water courses in the informal sector are largely missing. Data gaps make it a rather difficult proposition to develop a complete picture of water transportation in urban areas at the present juncture.

In a similar manner, since no conscious effort has been mobilised towards developing recreational areas with the help of water courses (except at places where such facilities are available traditionally), not much data and information are available on the contribution of such facilities towards improvement of the quality of urban living.

4.4   Planning Considerations

Certain studies have already been carried out to assess the potential of
the water transportation system in cities like Calcutta and Bombay.
Recently, a study has also been concluded for Delhi. These studies
have tried to focus attention on the navigational aspects, but, being
essentially studies in isolation, the findings of such studies have not
really been related to the overall transportation scenario and the land
use plans for such cities. These studies are generally silent on the
aspects of recreational facilities and water sports along such water
courses.

Economic viability is an important parameter for the introduction or
augmentation of any system, particularly when such a system has
been neglected for decades. Measures such as river management and
conservancy have been talked about but never really very seriously
attempted. Even now all efforts seem to get bogged down at the initial
stage itself in the controversy of high capital costs, associated with
high technology, or low operational returns, associated with the
conventional systems.

Being essentially a gift of nature, prospects of intra-urban water
transportation are highly location-specific. Interestingly, Calcutta
and Bombay being linear towns with spines along the north-south
axis, the major travel corridors are aligned in the same manner.
Studies indicate that for Bombay the eastern waters are safer for the
operation of even conventional crafts. The east-west cross corridor is
also considered safe. However, the west coast alignment is not
suitable for the operation of conventional crafts.

While on this, it must be mentioned that the planning criteria, within
the essential parameters of safety, efficiency, economy and environment, tend to be different for different categories of movement (e.g.
passengers in the form of commuters, goods generally in the form of
bulk commodities and people seeking recreational facilities). Low
travel time and a low travel cost will be desirable from the point of
view of commuter movement. But a low unit cost of transportation

will be desirable for movement of goods even with a higher travel time. A low travel time and higher travel cost may be acceptable to people seeking recreational facilities and to water sport lovers, provided other amenities are made attractive enough for this category of users. Such requirements tend to make competing demands.

Associated with this are the demands for landing facilities in each system. Considering the fact that commuters would prefer to use water based system only when the point of origin and destination are within the immediate vicinity of landing sites, requirement of transit facilities may not become an overriding factor. For goods traffic however, a *roll off-roll on* system would be very desirable. This will also dictate the type of crafts that must be used for the movement of goods along the water based system.

Goa has a record of movement of ores along the rivers. The crafts utilised for this purpose are, obviously, not adequately suitable for movement of commuters (for whom a 'water-bus' is needed). Goa also has tourist crafts and the demands for services of this nature are characteristically different from those required for the movement of goods or commuters. Thus, there is a need for cruise boats. Marine products is a major commodity transported on the water based system in Cochin. The nature of commodity demands an efficient system of transfer, from a water-based to a road-based transportation system, with landing facilities that could cater to such activities efficiently.

There is a great potential for water transportation in Madras as the water courses fan through the city. Delhi is a ring-radial city with the Yamuna aligned in the north-south direction. Incidentally, the north-south travel corridors are heavily congested in the case of Delhi also. Being parallel to the Ring Road on the eastern tangent, the river has a great potential of carrying a substantial volume of traffic (may be predominantly goods). In Calcutta services are already available along and across the Hooghly.

But, in all such cases, capital investment having been of a low order, ancilliary and support infrastructure could not develop to the desirable level, with the result that the quality of service has continued to remain

640

trapped in the vicious circle of the poor input-poor output mechanism.

If water transport for intra-urban travel is to be made economically viable and socially acceptable, intensive investments in this sector over a reasonably short period of time will be called for. The west coast waters provide a parallel corridor to one of the busiest transport corridors in Bombay. From an operational point of view, this water-course could be safely negotiable only for hovercrafts and hydrofoils. Out of these two, hovercrafts will be technically more suitable for this area. Practically, on all other intra-urban waterways, suitably modernised conventional crafts would prove to be adequate at the present juncture. Even for intra-urban waterways in the north-eastern parts of the country (e.g. Gauhati) similar systems could be adopted.

4.5   Recommendations

India, being endowed with a large number of perennial rivers and water courses, it would only be natural to explore and exploit the potentials of this system to meet the intra-urban travel requirements and to develop recreational areas that could go a long way towards improving the quality of urban living. The cities and towns where necessary studies, as mentioned elsewhere, could be initiated on a priority basis are:

- Calcutta
- Madras
- Goa
- Bombay
- Cochin
- Gauhati and
- Delhi.

The activities to be immediately taken up, in the context of introduction of intra-urban water transportation system, would be:

- to establish the role of water transport vis-a-vis the quality of urban living (e.g. passenger transportation, goods transportation, recreational etc.)

● to determine the traffic potential of each corridor (water based) in relation to the overall transportation scenario for an urban area (developed on the basis of present and proposed land use) - in fact, conscious decisions could be taken in locating various goods and commuter activities keeping in view the available waterway as a transportation corridor (thereby bringing down the apparently inevitable congestion levels on the road/rail network)

● to make a realistic assessment of engineering, management and maintenance costs associated with this system after carefully studying the physical and navigational constraints, the possibilities of overcoming them, the requirements of landing facilities and the facilities for modal transfer

● to examine the economic viability of the system (keeping in view the various types of subsidies available with the transport sector)

● to work out the priorities of a water transportation system in different urban areas, and on different stretches of water courses within the same urban area

● to draw up a judicious investment plan based on the findings of the studies mentioned earlier

● to create an agency under a unified authority that could be made responsible for the implementation of the system, wherever found viable

● to bring about closer interaction between user agencies and R&D units in this sector

● to initiate river front development, and the development of water bodies for recreational purposes, in any urban area as a mandatory exercise in land use planning.

Most of these cities are state capitals or important towns in their respective states. It is natural therefore that the developmental programmes are entrusted to states, with appropriate funds being made available by the Central Government as loans/grants. The R&D problems vary from city to city and can be best handled by the State Councils of Science and Technology.

5   Tourism

5.1   Background

The main objectives in the promotion of tourism are generally stated to be the following:

● Exploiting the tourism potential to support local handicrafts and other creative arts, and thereby creating large employment opportunities in the secondary and tertiary sectors of the economy

● Faster development of tourism, to attract foreign tourists, and thereby to generate foreign exchange

● Redefining the role of public and private sectors, to ensure that the private sector investment is encouraged in mainly developing tourism and the public sector investment is focused mainly on the development of the support infrastructure

● Promoting national integration

● According the status of industry to tourism.

The Central Government has been propounding a target of 2.5 million foreign tourist arrivals by 1990 based on the facts that one million tourists arrived in India in 1986. If this ambitious target is to be fulfilled (it is possible), more attention needs to be given to the development of tourism connected with water transport. It would be worthwhile to look at Table 7, showing the foreign tourist arrival to India during the last five years, and appreciate what massive efforts are called for if we are to reach the target of 2.5 million by 1990.

Table 7

The Number of Foreign Tourists during 1982-86*

1982	8,62,174
1983	8,84,731
1984	8,52,503
1985	8,36,908
1986	10,80,050

*Excluding tourists from Pakistan and Bangladesh.

In addition to attracting the foreign tourist to India, the local tourism also needs to be developed and there are indications that it is bound to expand due to:

- Increase in population
- Better standard of living
- Better purchasing power
- Tendency to spend for better living
- Desire to seek relief from the mechanical life in modern metropolitan cities.

It is a well-established fact that water in the form of sea, lakes and rivers holds out a natural attraction for tourists. This fact needs to be capitalised on in the general objectives of developing tourism in India.

5.2   Recommendations

If tourism is to be developed, we need to develop good transport systems, good roadways, good communication and good waterways and hotels etc. This is a case of inter-sectoral development with the principal goal of developing tourism. The entire western coast from Surat to Cochin can turn into an international sea-borne tourist zone provided the waterways are exploited and other necessary transport communication facilities are provided.

For developing tourism connected with water transport, the following points are of immense importance:

● In as much as roads need to be built, the navigable waterways also need to be provided and, for this, certain capital dredging and maintenance dredging is required to be carried out. There are lots of reefs, particularly on the western coast where the depths of the sea are tapering down in a gradual manner. There are also sand bars near the coast

● The selection of crafts is equally important if we are to assure quick transport to the tourist, to and from the metropolitan towns/ cities. Hydrojets and hovercrafts are very suitable means of transport. In the past these crafts used aviation fuel but, with the latest technology, they now use normal marine diesel and are no longer very expensive to run. In India we have, by and large, never thought beyond normal conventional boats, as sea-borne passenger crafts with 10/12 knots speed, when, in other countries, hovercrafts; hydrojets, with average speeds of 35/40 knots, have been used commonly for the past two decades. For example, a fast and fuel efficient ship between Bombay and Goa would bring about an immediate improvement in the tourist trade of Goa and its overall economy. Since the hydrojets and hovercrafts are small sea-going crafts, their manufacture should be considered indigenously with the help of transfer of foreign technology. Some of the existing public/private ship yards can also undertake the manufacture of such marine crafts with various capacities

The maintenance facilities for these marine crafts also need to be provided for their efficient use.

● The *shikara* concept of holidaying facilities with modern lodging and boarding amenities need not be the monopoly of the Kashmir Valley alone. Such *shikara* facilities can also be provided at every point of tourist interest on the western coast, particularly at Bankot Creek, Savitri, and the Vaishisthi rivers in Maharashtra

● As regards the availability of food in various hotels, particularly in *shikaras* and other type of hotels, it would be essential to develop the infrastructure for providing standard Indian food to the foreigners when they visit India, in addition to the food to which they are accustomed

● Various water sports can be developed with the opening of tourist spots on the banks of rivers and creeks and small harbours. The Government of Maharashtra has recently declared several areas near the waterways as special development spots for tourism. On the west coast, Goa and Cochin provide ideal locations for water sports. Wherever water sports are to be developed, by both in the private and public sectors, at such centres the import of implements needed for certain water sports may be allowed with duty exemption, and the equipment for some of the water sports can then be manufactured locally

● There are several potential spots of tourist interest away from the metropolitan towns, but they are neglected today largely because of transport bottlenecks. Several tourist spots on the western coast near the virgin sea beaches could be developed by improving access to them

● The development of various sites on various creeks and rivers, with large potential for tourist interest, should be left to individual State Governments with a time frame for the development of tourist centres. The foreign exchange accruing from tourist centres on the waterways would be enormous and would justify the initial investment

● It is desirable to acquire experience in this emerging area by initiating pilot projects to try out new kinds of crafts and develop new techniques for water sports. Pilot projects could investigate the impact on the congestion of road transport, the improvement in accessibility of remote areas, the environmental impact on banks, small vessels, marine life etc, transfer of technology and employment opportunities. The states of Goa and Kerala may be supported for such well-focused pilot projects.

6   Research and Development Requirements for Water Transport

6.1   Introduction

Before deciding on the thrust areas, for research and development in the field of water transport, it would be worthwhile examining the present-day constraints in the operation of water transport, the technological gaps that exist today and the needs of the country in this field. In this context it is necessary to ascertain the adequacy of channel dimensions vis-a-vis navigation of vessels, the stability of banks, sedimentation and dredging, the need to undertake excavation in soil, rock blasting, disposal of dredged material, river training works, estuarine works at suitable locations, jetties/landing facilities, locks, gate designs etc. In view of the inter-relationship of many of these factors, the advice of experts in areas such as navigation, ship hydrodynamics, soil and rock mechanics, dredging, environmental pollution etc. is essential. An added problem in some of the waterways, like those in Kerala, is the abundant growth of water hyacinth, whose removal also requires expert guidance from agronomists and botanists.

The major constraint to the smooth operation of inland navigation canals is the non-availability of adequate depths. For safe and organised navigation, a sufficient depth of water throughout the year and a fairly low velocity of flow are essential. Increasing water withdrawal for irrigation and the lowering of the water table due to adverse environmental conditions, caused by indiscriminate deforestation, pose a continuous threat to the already critical depth available in the waterways.

While considering the development of water transport systems in Indian rivers the characteristic features of these rivers should be kept in view. The flashy nature of a river, with discharge limited to only a few months in a year, the existence of hard rock at shallow depths in the upper areas, the tidal influence near the mouth, the meandering pattern in different regions, the large fresh water discharges, the existence of braided channels and large sediment movement are some of the features that we have to contend with.

Severe erosion along the banks, with consequent siltation of channels of the river, is another important problem requiring attention. Constant maintenance dredging becomes essential in many reaches of the river to maintain adequate depths for navigation.

In the case of water transport along the sea routes, maintenance of adequate depths at the landing terminals and over the bars on the way is necessary. Providing adequate shelter at the terminals against wave action is also very important.

In the case of canals, erosion along the banks due to waves generated by ship motion is a major constraint requiring careful attention. The basic nature of soil comprising the bank must be examined before suggesting any protection measures. This calls for studies in the laboratory, simulation of the movement of vessels and of the likely wave and current forces acting on the banks. The stability of the suggested protection methods also needs to be checked with the help of flume studies.

The dimensions of a waterway dictate the maximum possible speed that could be achieved by a vessel. It is well established that when the Froude number (the ratio of v to the square root of the product of g and d; v=speed of the vessel, d=depthof the canal and g=acceleration due to the gravity) approaches the limiting value, the power requirement of vessels increases considerably, and beyond the critical value of Froude number equal to one (for alongside flow), any increase in the power input does not increase the speed of the vessels. Depending on the dimensions of the channel and the vessel to be used, the blockage ratio would determine the value of the Froude number likely to be encountered.

Depending on the size of the vessel the waterway has to be designed properly, especially at bends, to provide the required radius of curvature for smooth navigation. This requires interaction between the navigation specialist, the hydraulic engineer and vessel designers, particularly in the absence of established practice. Several investigations need to be carried out to determine the behaviour of a vessel under such circumstances as well as the effect of flow conditions resulting therefrom on the stability of the channel.

The problem gets more complicated when one proceeds down the waterway to the reach where the interaction of the tidal and fresh water flows begins to be felt. The navigational aspects are modified by the tidal levels which vary with time. The problem of sedimentation in the navigation channel could become complex due to the interaction of salt and fresh water leading to partial or no mixing and the occurrence of a salt water wedge. The sediment movement is then controlled by the movement of the salt water wedge whose position also varies depending upon the quantum of fresh water flows and tidal discharge. The effect of deepening the navigation channel in such cases has to be carefully considered since this could completely offset the expected improvement, as the salt water wedge could travel further upstream after deepening, causing greater sedimentation problems. The behaviour and stability of coastal inlets also becomes an important aspect of study in certain cases. The problem becomes more complex where a large amount of littoral drift is encountered at the mouth of the inlet which needs to be kept open against this massive sand flow. Besides the stability of the vessel, the entry at the inlet could be affected by the formation of a bar at the inlet and the consequential action of breaking waves over the bar.

Dredging is a common feature in all waterways and disposal of the dredged soil needs to be carefully planned to avoid not only the problems of reshoaling, but also environmental pollution. A careful study is therefore essential to ensure that no damage to the environment results from such disposal.

Locks are provided through inland waterways to overcome the change in the variation in water level caused by abrupt changes of bed slope. The design and operation of such locks need model studies, with suitable instrumentation, to ensure that hydraulic forces on the vessel do not exceed permissible limits.

Excessive aquatic weed growth results in swamp formation and changes in hydrochemistry, and thereby reduces fish production. Aquatic weeds reduce the productive characteristics of transport canals, reservoirs and cooling ponds. They also endanger swimmers, foul boat propellers and make certain types of fishing activity impossible. They can reduce water flow in irrigation ditches, and result in large transpiration rates. In large waterways, that are customarily used by boats, weeds can choke the channel to the point where it is impossible to get a boat through.

Excessive inorganic nutrients (phosphates, nitrates, silicates etc.) can cause eutrophication of lakes and fertilizing flowing streams, with the resultant production of heavy plankton blooms. Water plants such as water hyacinth, water milfoil and water chestnut can clog water channels and interfere with various recreational and navigational uses of water. Algae, worms, barnacles etc. can clog passage ways, cling to vessels and accelerate corrosion. Therefore, the control of aquatic weeds with the help of expertise in agronomy and botany is essential.

6.2   Thrust Areas for R&D Efforts

From the above it would be apparent that, in order to achieve significant development in the field of water transport, a significant supporting R&D effort would be required in the following six important areas: (a) coastal and port development (b) inland water transport (c) shipping and ship building (d) conservation (e) environment and (f) geotechnical and structural aspects. The possible R&D needs on each of the above are given below:

Coastal and Port Development

India has a large coastline of about 5700 km length and there are 11 major ports and around 150 small and intermediate ports. The traffic handled by these ports is about 130 million tonnes per year. The important problems which need to be looked into, in the case of coastal and port development to improve coastal shipping, are the littoral movement and the consequent erosion and accretion, coastal defence, stabilisation of inlets, dredging and disposal etc. The thrust areas in this field are given below:

● Study of littoral processes along the east and west coasts of India. Estimation of littoral drift through field studies, observations on waves, wave induced currents, beach profiles, siltation and scouring pattern, sand bypassing, beach nourishment, sediment budgeting, tracer techniques, mathematical modelling, coastal zone management

- Behaviour of coastal inlets along the two coasts of India, evolvin stability criteria for inlets
- Ship motion studies, navigational problems, ship manoeuvring problems
- Erosion and siltation - assessment and control
- Materials for construction and for use in manufacture of marine equipment
- Design of coastal defence works and other harbour/offshore structures
- Pollution control and abatement
- Improving the utilisation of specialised equipment
- Instrumentation for obtaining design data.

Inland Water Transport

In order to make water transport economically viable and competitive, compared to other modes of transport, it is necessary to concentrate on factors which are responsible for the extremely slow development in the field of inland water transport in India. The principal factors are the use of outdated and inefficient vessels, lack of terminal facilities for loading/unloading, inadequate depths due to lack of conservancy measures, deterioration of rivers and canals and the non-availability of navigational aids. The role of water transport in the country has, therefore, been negligible so far. The thrust areas are given below:

- Design of terminal facilities for inland canals
- Design of canals, canal linings, bank stability
- Design of dredging equipment suitable for inland canals
- Sedimentation problems in inland canals
- Design of suitable navigation locks ·
- Development of means and material for keeping the national waterways and depths of water in proper shape
- Development of the design and material for the construction of country boats which are most suitable for the inland water transport system

● Design of equipment for efficient removal of weed growth.

Shipping and Ship Building

There is a need in India not only to improve the inland and coastal waterways for navigation, but also to modernise the vessels plying in the national waterways, to give them a proper and safe hydrodynamic shape, to make them most efficient in respect of fuel consumption, manoeuvrability and safety. Given this background, a thorough study of the inland waterways, and the vessels operating in them, on scientific lines is called for. The thrust areas in this field are:

● Modernising the vessels to make them safe as well as fuel efficient
● Development of a safer and more economic model of a ferry system with high speed catamarans, multi hull or SWATH ships (small waterplane area twin hull ships)
● Manoeuvring under the effect of currents, waves, winds at bands, crossings etc.
● Development of coastal shipping, development of suitable terminals for passenger/ferry service
● To development of standard computerised designs for ships, trawlers and other vessels of standard sizes to reduce the problem of spares
● Indigenisation of consumables and other equipment used for the industry
● Development of suitable designs for hull etc. which may reduce fuel consumption.

Two models of vessels designed for specific uses are described in Appendix-2 and Appendix-3. These are good examples of what can be achieved through R&D effort.

Conservation

One of the objectives of the Seventh Five Year Plan was the implementation of comprehensive conservancy works for each waterway and the preparation of a systematic plan for conservancy works, including hydrographic surveys, dredging, maintenance of channels and setting up of navigational aids. A separate dredging unit was also considered to undertake capital and maintenance dredging to maintain navigable depths. The following R&D studies need to be considered.

- Design of river improvement works, river bank protection
- Stabilisation of river bed and banks
- Design of dredging equipment suitable for navigable rivers and canals
- Dredging in canals and the disposal of dredged material, and its use for reclamation
- Ship generated waves and the design of canal linings, low cost methods of canal lining
- Controlled blasting in canals
- Estimation of siltation in canals connected to rivers or creeks and the need for training measures
- Seepage through canal banks and its effect on bank stability.

Environmental Aspects

The following areas require a research effort:

- Water quality problems due to dredging and disposal of the dredged material in rivers/canals
- Aquatic weed growth in rivers, canals, reservoirs and methods to control and remove the growth
- Problems related to the industrial/municipal/thermal/agricultural effluents in rivers/canals and methods to minimise its impact on fish life and other benthic organisms
- Problems related to the increase in the inflow of saline waters as a result of deepening of channels in rivers, estuaries etc.

Geotechnical and Structural Aspects

The thrust areas in the field of geotechnical and structural aspects on water transport are:

- Bank stability under action of ship generated waves and currents

- Design of pitching/lining and slopes for hydraulic conditions,and against slip failure
- Low cost methods for canal lining design
- Dredgeability of soil and hard strata, safe distances for rock blasting etc
- Seepage flows and their effects on bank stability
- Impact forces of barges on bridge piers.

6.3   Suggested Action Plans and Pilot Projects

The following schemes are suggested as action plans or pilot projects, to be taken up under funding and monitoring by the Central and State Governments:

- Creation of an efficient and extensive database, at a convenient location, comprising:

  - Detailed maps of navigable rivers, giving water levels, channels, drafts
  - Detailed drawings of port and navigation facilities, constraints, due to bridges, other structures and natural formations, for inland water transport and coastal water transport along specified waterways
  - Time series data on traffic catered by inland water transport and coastal water transport
  - Data on craft and vessels
  - Data on manpower, expenditure, plan targets and achievements.

- Transport demand assessment and forecast for the years 2000 and 2010, with estimates of potential diversion of traffic from road and rail to water transport
- Techno-economic feasibility studies for establishing and augmenting facilities in selected inland water transport systems:

  - Zuari and Mandovi rivers
  - Ganga-Bhagirathi Hoogly system
  - Brahmaputra-Bharak system

O   West Coast Canal and Kerala back waters
O   Brahmani-Mahanadi system
O   Godavari-Krishna system
O   Narmada river system.

The study in each case should include the traffic forecast, location of terminals, availability of other infrastructure, and the estimated costs and benefits.

● Techno-economic feasibility studies for establishing/augmenting facilities for coastal shipping along selected routes:

O   Madras-Tuticorin coastal route, through Pamban and Palk Strait
O   Madras to Port Blair (augmentation)
O   Tuticorin to Great Nicobar
O   Great Nicobar to North Andaman servicing the island in between
O   Cochin to Lakshadweep
O   Bombay-Goa (augmentation)
O   Jafrabad to Bombay/Mangalore. Further to the provisions of SCI proposal, the additional infrastructure required in terms of port improvements, interface between port and road/rail modes of transport, development of hinterland and availability of articulated trucks for use with Ro-Ro service may be studied. The feasibility of extending the service to Cochin may also be examined.

● Study of techno-economic features and defence implications, of the overall development of the Great Nicobar Island, with coastal shipping as a major component (including establishment of a ship repair yard).
● Research studies towards the development of suitable vessels for inland water transport and coastal shipping, including design of suitable engines and on-board handling gear.
● Development of guidelines for design, operation and maintenance of inland waterway terminals, coastal shipping ports and ports handling both inland waterway vessels and coastal shipping

vessels, with regard to berths, passenger movements, cargo handling, warehouses, intermodal transfer facilities and other provisions. The possibility of developing joint facilities for coastal shipping and fishing may also merit study.

- Research on channel and river conservancy measures for the selected waterways. Standards are to be evolved for optimum cross sections of navigable channels, vertical clearance and lining/embankments.

- Examination of an integrated approach to regional development planning with water transport as a major constituent.

- Study of recreational uses of inland waterways.

- Feasibility studies to establish a national institute, with regional centres, for training personnel at the various levels of operation and management of inland water transport. The maintenance of databases could be made the responsibility of this institute.

- In order to implement the research programmes it will be necessary to develop manpower in the water transport sector and to identify the research organisations and educational institutions in the country which will be able to conduct the basic and applied research required in this respect. In Appendix-4 we list centres within the country which can contribute solutions to the various problems described above. A coordinated framework is needed that brings together the users, the R&D centres, the funding agencies and the production units.

Water transport is also a multidisciplinary activity and expert advice would therefore be required in the field of navigation ship hydrodynamics, soil and rock mechanics, coastal hydraulics, dredging, environmental pollution etc. A central organisation such as the Inland Water Authority of India should be in a position to coordinate these activities for IWT while the Shipping Corporation of India could do likewise for coastal transport.

● A comprehensive programme is recommended to collect coastal data in respect of waves, winds, currents, siltation and scouring littoral drift, behaviour of tidal inlets etc. along the two coasts of India at selected locations. A data bank may be set up where these data would be stored, processed and retrieved when required.

● The Ship Hydrodynamic Research Centre, which has been proposed by the Central Water and Power Research Station (CWPRS), may be set up to study the ship design and navigational safety aspects of water transport.

● A separate research cell may be set up to study the dredging technology required for inland navigation. This could be attached to a hydraulic research station, such as, the CWPRS, and the DCI could be associated with such a cell.

● A research cell to study the environmental aspects of water transport may be set up. The cell may be attached to an appropriate research organisation in the country working in this field.

● The State Councils of Science and Technology should actively participate in the above R & D projects pertaining to their local regions.

7   Recommendations

Some of the detailed recommendations have already been listed at the end of each section. The *major* recommendations, and the action points, are listed in this concluding session.

● Coastal shipping has considerable potential for transporting cargo like crude oil, petroleum products, liquified natural gas, thermal coal etc. Planning at the national level should take this into account.

● Passenger services along the West coast and to and fro services to the Andaman & Nicobar as well as Lakshadweep islands have to be developed, as part of a national strategy of development of these areas.

● The country needs to augment its effort in cargo transport by resorting to containerisation to compete with similar developments internationally.

● An integrated approach to inland water transport needs to be followed by developing national waterways. This involves providing an assured navigation channel, modern cargo handling terminals and linkages with other forms of surface transport. Central and State Governments and private agencies should participate in this task.

● The State Governments may be encouraged, with 100 per cent loan assistance, to take up techno-economic studies and the improvement of waterways in the states.

● The Inland Water Authority of India has to play a lead role in the R&D infrastructure of the IWT. This can be made possible only if the IWAI is given sufficient autonomy to operate all its programmes.

● We have identified a number of cities where the traffic congestion on roads can be considerably relieved by opening suitable corridors of urban water transport. The concerne' State Governments should take the initiative to carry out feasibility studies and they could be further supported in planning by loans from the Central Government.

● To attract tourists and foreign exchange, water sports, cruises and boatels (hotels on water) can be developed in places which are suitably located on the coasts, creeks, lakes and river banks. The State Government tourism departments should be the nodal agencies for this activity, although private enterprise can also contribute its share.

● Water transport development requires R&D studies in several areas, including coastal and port development, solutions of problems of IWT, shipping and ship building, conservation, environmental problems and geotechnical and structural aspects. We have outlined several action plans and pilot projects, and indicated the R&D institutions in the country where these can be carried out.

● Finally We strongly feel that, with the modern inputs of science and technology indicated in this report, it is now the right time for dedicated planning, to tap the tremendous latent advantages that water transport can provide.

Appendix-1

Strategic Aspects

Waterway linkages to the north-east

The rivers Brahmaputra and Barak, along with their tributaries, are two of the most important rivers of the north-east. Both these rivers are connected to the Ganga only through Bangladesh, where navigational rights are to be extended by a yearly protocol between the Governments of India and Bangladesh. Vital supplies like foodgrains, construction material are being transported today by CIWTC to the north-east through Bangladesh. Traffic moving through Barak river terminates at Kharimganj which is the nodal point for the north-eastern states of Tripura, Manipur, Mizoram, Assam and Nagaland. There are plans to develop an integrated water management system in the Barak valley giving due importance to navigation. A major dam is planned at Tipaimukh on river Barak by which the navigation from Kharimganj could be extended upto Tipaimukh, a distance of about 250 km.

Whereas the Brahmaputra is a mighty river which is already navigable, the critical factor, for assured navigation links to the north-east, continues to be the linkage through Bangladesh. While it is not feasible to link Barak with Ganga through Indian territory, it is feasible to link Ganga and Brahmaputra through Indian territory. At the same time, the Barak river, which is called Kushyara in the Bangladesh territory, is very shallow permitting a depth of only 1.8 m due to rock outcrops in the river bed. While the dependence on Bangladesh for our transport requirements to Kharimganj remain, it would be necessary to explore the possibility of deepening and widening the navigable channel in the Kushyara river to at least a 3 m depth of water, so that vessels of draft 2.4 m could be operated between Calcutta and Kharimganj for economy of transportation.

On the other hand, it is feasible to link up, the Brahmaputra and the Ganga either through Bangladesh or through the Indian territory. The latter alternative is preferable as this third mode of transport will guarentee uninterrupted communication between the north-east and the main land. It is understood that the link canal, of approximately 450 km, would be a capital-intensive scheme. The Ministry of Water Resources may,however, re-examine the necessity, in the context of strategy, while, at the same time, ensuring a transport linkage between the Ganga and the Brahmaputra (which are being given a greater priority for the development of national waterways). It is needless to mention that such a canal could serve other purposes like irrigation; so its viability would, really be much more broad based.

Coastal Shipping between Madras and Tuticorin

Currently, we do not have a reliable coastal shipping route between Madras and Tuticorin suitable for the operation of moderate-sized ships. Such a facility is essential for the movement of coal for the super-thermal power station at Tuticorin, and for the economic development of the region comprising the southern tip of the Indian peninsula. Further, coastal shipping from the ports on the west coast have to sail around Sri Lanka to reach the ports on the east coast. Such navigation entails a longer sailing time and additional fuel consumption making the shipping mode, less cost-effective and exposing shipping to additional hazards.

By dredging a shipping channel, and building in more navigational infra-structure in the area near Palk Strait and Pamban, it is possible to provide a permanent sea-way within the economic zone of the Indian peninsula.

The economic benefits of such a project are obvious. The development of Tuticorin port into a major international port will be accelerated, with spin-off benefits of economic prosperity to the hinterland.

The desirability of navigating coastal shipping within the economic zone of the country cannot be over-emphasised. The development of the coastal route between Madras and Tuticorin will aid in the eventual development of naval facilities in Tuticorin, for surveillance of the seas around the southern part of the peninsula along with the facilities in Cochin.

The proposed shipping route between Madras and Tuticorin will reduce operating costs and will also provide a more convenient shipping connec-tion between Colombo in Sri Lanka and the Indian ports on the east coast. Besides stimulating trade between the two countries, the proposed route will also facilitate greater international cooperation in other spheres of mutual interest.

Appendix-2

A Seabus System
Extracts from reports of S.H.M. Marine
International Inc.

Columbia and the City of Vancouver is on the south shore of Burrard Inlet. Two crossings were in existence over this stretch of water: the Lions Gate Bridge and the Second Narrow Bridge. The north and south approaches to both bridges were beyond the traffic saturation point and a third method of crossing was mandatory.

In 1974 the Bureau of Transit Services (as it then was called) of the Ministry of Municipal Affairs of the Province of British Columbia decided to establish, as the third crossing, an efficient passengers-only ferry link between the foot of Lonsdale Street in North Vancouver and the most convenient location in the vicinity of Granville Square on the south shore. This ferry link would form an integral part of an overall public mass transit system arranged with convenient bus/ferry transfer points and a common through fare system. This would alleviate the bridge traffic problem by reducing the number of buses and private passenger cars and would ease the downtown parking problem.

In June 1977, the Seabus system went into operation with two vessels providing service across Burrard Inlet between Vancouver, B.C. and North Vancouver. (North Vancouver is on the North shore of Burrard Inlet).

The components of the Burrard Inlet Seabus System are the ferries, the floating passenger terminals, the north shore bus/ferry transfer points, the south shore overhead walkway to the bus transfer or dispersal point, the floating servicing berths and the shore based maintenance and administration building. The system can also, if required in future, interface with an LRT system or other ferry systems. The original-two ferry system can be expanded to eight ferries without additional passenger terminal facilities.

The marine link has a crossing length of 1.75 nautical miles and design criteria were set for a departure from each terminal of one ferry every 10 minutes. Consequently, a double-ended catamaran configuraton was selected for the vessels: catamaran to provide good resistance to heeling while loading/unloading at the terminals; double-ending to elimate the dangers and time loss in turnaround at the terminals.

One of the important considerations in setting up an economical operation is fuel efficiency. Minimisation of displacement, service speed and idle time in the terminals, all consistent with peak hourly passenger capacity requirements, contribute significantly to fuel efficiency. To this end, aluminum was selected as the hull construction material because of its lighter weight and also because of its ease of maintenance. A service speed of 12.5 to 13.0 knots gave a terminal to terminal time of nine minutes leaving one minute for loading and unloading.

In order to achieve the desired in-terminal time a flow-through simultaneous loading/unloading system was adopted. The interior passenger space onboard the ferry was arranged to form six similar rectangular areas each served by a double door at either side to match an equivalent flow pattern on the terminals.

The adherence to a tight schedule, when operating in a busy waterway, required good manoeuvrability. The double-ended configuration dictated that the manoeuvrability and speed be available in both directions. The answer was provided by installing four steerable Z-drive units. With these the ship can be moved forward, backwards, sideways, diagonally or crabwise and it is possible to bring the ship to a full stop in the water from full speed ahead in about one ship's length. Besides, either end can readily become the bow or the stern at will.

A schedule that allows one minute to arrive, unload, load and depart demands that the terminals be designed to adjust automatically to tide changes and be arranged to suit the flow-through the loading/unloading system. Furthermore the terminals would have to provide for future expansion of the ferry fleet.

For each end of the crossing a floating passenger terminal with the shape of an E would automatically compensate for the expected tide variation of 18 feet and would provide two berths. The centre finger, arranged as the loading side, and the outside fingers, arranged as the unloading sides, would provide the flowthrough system. Opposite the centre finger a shore connection would be provided by a transfer ramp hinged on the shore end and suitably railed to direct passengers in the desired direction.

The dimensions, shape, function and anchoring arrangement of the terminals preclude even drydocking them for maintenance or repairs. An investigation of the various options available for construction materials, and a comparison of their respective capital costs and maintenance requirements, indicated that prestressed, post-tensioned concrete would be the most suitable.

The inside faces of the berths are lined with a shock-absorbing fendering system, having a face-to-face dimension across the berth, which will provide the ferry with a neat fit in the terminal. This system coupled with the shock-absorbing buffers at the head of the berth ensure that the access doors, on the ferry and the terminal, line up to provide straight through access and exit.

A busy commuter ferry requires daily servicing for such things as janitorial work for passenger spaces, exterior washing, refuelling, waste liquid pump-out, and general servicing work. E shaped maintenance terminals built up from ferro-concrete pontoons are provided for this work and connections for the required services, including shore power connections, are provided.

A transit or transportation authority, who have to move masses of people across a stretch of water on a commuter basis, or an industrial or business area, separated from the domicile area by a river or lake, can profitably use a Seabus System. A Seabus System can turn a river or lake into an advantage where commuter travel along the bank could be water-borne rather than taking up road space.

Appendix-3

A Near-Shore Platform

The platform illustrated in the following drawings was designed at the Nationa Institute of Oceanography, Goa for use along the coastline of Maharashtra. It has many useful purposes some of which are described below (see Figures 1, 2 and 3).

● It would provide a status report on tne marine environment of Maharashtra coast which would be useful for future developmental programmes of the state for port development, fishing operations, fishery resources.

Fig. 1 Photograph of the near-shore platform which has been used for loading the ore. The platform can also be considered for cargo loading/unloading in waters of about 30 m depth.

● It would make available trained personnel and the necessary infrastructure for future environmental programmes of the Government.

● It would initiate a pollution monitoring programmme of the state which is necessary for the welfare of both the coastal population and the tourists.

● The data collected will provide the necessary information for marine culture thus enhancing fishery production as well as providing employment for the people of the coastal area.

● The information would prove beneficial for protection of timber resources by helping in the preservation of wooden hulls of fishing boats.

● If required, these platforms could help in the implementation of various legal measures introduced for fishery protection and pollution control.

● The platforms could also be incorporated in strategic defence operations

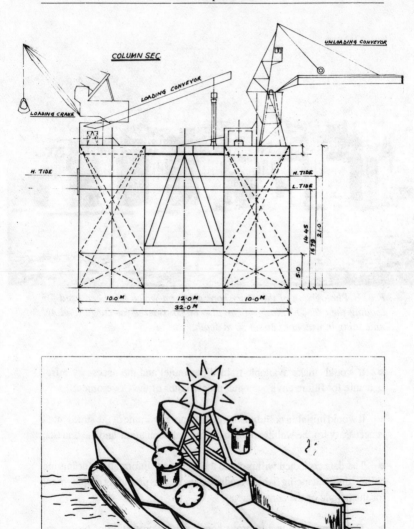

Fig. 2,3 The column section of the platform (top) and a small platform-cum-floating light vessel useful for passenger traffic. This vessel has been approved by the Port Authority of Maharashtra.

Appendix-4

Table 8 shows the R&D institutions concerned with problems relating to water transport.

Table 8
R&D institutions working on water transport

Organisation	Subject of R&D activity
IIT., Kharagpur	IWT vessels and ship model testing
IIT., Madras	IWT vessels and ship model testing
National Ship Design and Research Centre, Visakhapatnam (a new organisation in the offing)	IWT vessels, ship design
Central Water & Power Research	River Engineering, ship model Station, Pune testing, ecological problems.
Central Board of Irrigation and Power, New Delhi.	River engineering
National Remote Sensing Agency, Hyderabad	Remote sensing for hydrographic survey and river morphology.
Central Road Research Institute New Delhi.	Formation of roads on river beds
Large Public Sector Shipyards (Cochin Shipyard, Hindustan Shipyard, Mazagaon Dock Ltd. and Garden Reach Shipbuilders and Engineering Co.)	Commercial ships (first two) and defence production (last two); Coastal ships, IWT vessels and harbour crafts.
Medium size public sector/state Govt. owned shipyards (Goa shipyard, Rajabagan Shipyard	Harbour crafts, IWT vessels

Table 8 (*Contd.*)

Organisation	Subject of R&D activity
CIWTC, Hooghly Dock & Port Engineerings, Shalimar Works and Alcock Ashdown Co. Ltd.)	
National Institute of Oceanography, Goa.	Harbour crafts (see Appendix-3)
Cochin University of Science and Technology, Cochin.	Ecological studies.
Shipbuilding and Design Dept of Ministry of Surface Transport.	IWT vessels, ships
NATPAC, New Delhi and Trivandrum	IWT planning and infrastructure

14
Minerals Development

Everything has been thought of before. The
difficulty is to think of it again.

Goethe

Contents

Contents

14
Minerals Development

Preamble

This report on minerals development has been prepared in response
to the following two questions referred to the SAC-PM.

● Whether our mineral resources, particularly metallic, are likely to
become redundant in the future years as a consequence of the
development of advanced substitute materials? If such a possibility
exists, should we not export to a greater extent our mineral resources?

● What should be the elements of our country's mineral policy?

In order to achieve clarity in preparing the answers, it became
essential to collect quantitative data pertaining to availability of
mineral resources, rate of their production and consumption, quantities of imports and exports and such other related factors. An effort has
been made to collect relevant data but there may be some errors in the
numbers that appear in various tables of this report. Although this
word of caution needs to be uttered, it is emphasised that such errors
do not affect the analysis embodied in the report. Again, in the interest
of clarity, only metallic minerals have been dealt with, and among
these also, only a few have been chosen for detailed comment. The
principles of analysis equally apply to other metallic minerals, as well
as to non-metallic minerals for which only the relevant data have been
presented. Therefore, although the exercise is by no means exhaus-
tive, the central point of the analysis is that the minerals picture that

is drawn for the country should cover the entire range of activities: extending from exploration all the way to the finished products and to their optimal utilisation; only then could an integrated minerals policy be evolved which will be noteworthy in the overall strategy for industrial and economic development of the country. At the end of the report, a summary of recommendations is presented that highlights this approach, and also the need for a high power body to take major policy decisions and initiate necessary actions, in the context of the suggested approach, with respect not only to strategic minerals but also those that have a bearing on the emerging new materials and technologies. ■

1   Status of Metallic Mineral Resources in India

The periodic table of elements is shown in *Fig.1* in which metallic elements are highlighted. It is clear that more than 80% of the elements are metallic elements. The minerals corresponding to these will be the subject of our analysis.

The status with regard to mineral resources for various metals is shown in a compact form in Table 1. The number written alongside each metal is its atomic number (see Fig.1). In this table the resource position with respect to metallic minerals has been classified as

- abundant
- medium
- insufficient and
- not known.

The above categorisation is based on the definition that if the existing and proven mineral resources of a given metal are sufficient for about thirty years (given the present demand and its rate of increase), the same can be considered *medium*. If the resource position is more, in terms of this criterion, it is deemed *abundant*. If the resource of a metal cannot even last for thirty years, on the basis of resources presently known, *insufficient* is the category. The *not known* category corresponds to metals whose resource position is unknown possibly

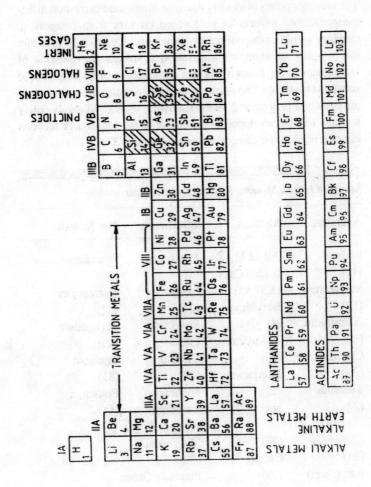

Fig.1 The periodic table (Source: C N R Rao, P P Edwards, Proc. Indian Acad, Sci., 1986). Hatched elements become metallic only in molten state.

because no serious effort has been undertaken as yet to identify their sources.

The minerals policy obviously has to be formulated in relation to this categorisation. Aspects to be focused on vary from category to category within an overall policy framework. For instance, it is only the category of abundant minerals that attracts the question of redundancy in the future years as a consequence of development of substitute materials. The issue of export again relates to only abundant minerals and is dealt with in the following section. Minerals policy features will be developed in the course of the subsequent sections keeping in view the categorisation presented in Table 1.

Table 1

Status of Indian Mineral Resources for Metals

Abundant	Medium	Insufficient	Not Known
Sodium (11)	Alkali Metals (Li,Rb,Cs)	Potassium (19)	Scandium (21)
Magnesium (12)	(3,37,55)	Cobalt (27)	Germanium (32)
Aluminium (13)	Beryllium (4)	Strontium (38)	Technetium (43)
Calcium (20)	Vanadium (23)	Yttrium (39)	Thallium (81)
Titanium (22)	Manganese (25)	Niobium-Tantalum (41) - (73)	Polonium (84)
Chromium (24)	Nickel (?) (28)	Molybdenum (42)	
Iron (Steel) (26)	Copper (29)	Platinum Group (Ru,Rh,Pd,Os,Ir,Pt) (44,45,46,76,77,78)	
Zinc-Lead (30) (82)		Silver (47)	
Gallium (31)		Cadmium (48)	
Zirconium-			

Table 1 (*Contd.*)

Abundant	Medium	Insufficient	Not Known
Hafnium		Indium	
(40) - (72)		(49)	
Barium		Tin	
(56)		(50)	
Light Rare		Antimony	
Earths		(51)	
(La, Ce,Pr,Nd)		Tellurium	
(57,58,59,60)		(52)	
Radium		Heavy Rare Earths	
(88)		(Pm TO Lu)	
		(61 TO 71)	
		Tungsten	
		(74)	
		Rhenium	
		(75)	
		Gold	
		(79)	
		Mercury	
		(80)	
		Bismuth	
		(83)	

Actindes (89 to 103) left out: numbers in brackets refer to atomic numbers.
? - Technology for extraction from low grade ores is recognised as a thrust area and is receiving serious attention.

2   Export of Metallic Minerals

We shall take up the general question of whether various metals are likely to be substituted. Substitution has certainly taken place to an extent, and this has had an impact on the metals industry (for example, see *Fig.*2 for the trend in per capita consumption of steel and aluminium in USA indicating saturation). An outstanding instance of

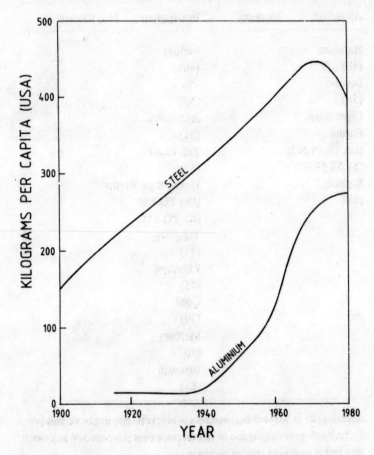

Fig.2 Per capita consumption of steel and aluminium in USA. (Source: Scientific American Vol. 254, No.6, 1986).

substitution of a fairly common metal is the substitution of copper by optical fibres in communication systems. Similarly, plastics have replaced steel and substituted for zinc in consumer related goods. Fibre reinforced plastics have displaced steel and, especially, aluminium alloys in certain structural applications. Ceramics are competing with metals and alloys for automobile and aircraft engine components. However, we seriously doubt whether complete substitution

Table 2

Basic Mechanical Properties of Materials

	Yield Strength (MPa)	Elastic Modulus (GPa)	Fractures Toughness (MPa m)
Metals and Alloys	20-2000	10-500	6-250
Structural Ceramics	1000-10000	100-1000	0.5-10
Engineering Polymers	7-120	0.1 - 10	0.4 - 7

will occur in the foreseeable future. This statement is supported by Table 2 in which is presented the mechanical properties of the three basic groups of materials, namely metals, polymers and ceramics, are presented. The contents of this table indicate that, among the three basic types of materials, metals possess an unsurpassed *combination* of yield strength (a measure of resistance to deformation), elastic modulus (a measure of stiffness) and toughness (a measure of resistance to fracture). Because of this, in the area of engineering structures, where components have to be designed to withstand impact or oscillatory loads or operate at cryogenic temperatures, for which reasonable levels of fracture toughness are desired, we have very little choice but to use metals. In the area of high temperature service, the competition can be only between metals and ceramics and again the inherent ductility of metals and alloys places them at a distinctly greater advantage. Considerations of reliability and long-term durability also favour metals, at this juncture, for a number of large tonnage structural applications. Apart from the fact that these characteristics enable metals and alloys to perform in a way that the advanced non-metallic materials perhaps cannot, cost and energy considerations favour metals and alloys, especially steels (see Table 3 & 4). Furthermore it is important to recall that metals dominate the periodic table of chemical elements and therefore, while the consumption of some metals may be showing downward trend, there will always be others with expanding use. For instance, the present trends suggest a recovery for non-ferrous base metals and increasing

Table 3

Comparison of Cost of Materials

Materials	Basic Cost* (£/Tonne)
Metals	
Mild Steel	210
Low Alloy Steel	400
Copper	1200
Aluminium Alloy	1100
304 Stainless Steel	1250
Titanium	5000
Plastics	
PVC	1900
Polythene	2000
Nylon	5250
Ceramics	
Silicon Carbide	275
Alumina	650
Magnesia	950
Composites	
CFRP	90,000
Boron-Epoxy	150,000

* Basic cost refers to UK , 1980: *Engineering Materials* by M F Ashby, D R H Jones Pergamon 1980.

attention to esoteric metals and their recovery (gallium from Bayer liquors, germanium by solvent extraction, selenium and tellurium from copper anode slimes and indium from other non-ferrous metal circuits). Even in the case of steels where a downward trend in terms of volume of consumption is to be expected and observed, it should be noted that there has been a marked substitution of some of the

Table 4

Cost and Energy Per Unit Strength for Steels and Plastics

Material	Kilogramme Cost (£) Per Unit Strength (x10⁻³)	Total Energy (kwh) Per Unit Strength
Metals		
Mild Steel	0.8	125
Low Alloy Steel	0.4	125
Cast Iron	1.2	70 - 300
304 Stainless Steel	2.5	230
Plastics		
PVC	38.0	800
HDPE	55.0	930
Nylon	61.0	600
Polythene	25.0	1555

Source: *Metals and Materials*, November 1981.

earlier varieties with newer steels which have improved quality and enhanced specific strength (e.g. high strength low alloy steels in place of mild steel) which can perform equally well at lower weights in the same type of applications.

Thus, no wholesale substitution of metallic materials in large engineering structures (which consume a considerable volume of materials) by polymers and ceramics is expected for at least a few decades. However, we should note the considerable progress in research and development aimed at overcoming the present drawbacks of non-metallics, e.g. modulus of polymers and toughness of ceramics, and that, as a result, the domain of application of metals will certainly shrink with time. Innovations in design which will allow the use of polymers, ceramics and composites will also become more prominent in the coming decades (M.F. Ashby, *Acta Metallurgica*, Vol.37, 1989, p.1273) leading to their increased usage. Metal matrix composites, which have a metal as the base and a non-metal as the reinforcement, are promising to be the material of the future and their emergence will ensure an even more lasting place for metallic materials in the far future.

Notwithstanding what has been stated above, it is necessary to consider the availability of mineral resources, their projected demand and the current levels of export before a view can be taken in regard to either augmentation or reduction in their exports. The data pertaining to the export of metallic minerals are presented in Table 5. The value of our country's export in 1985-86 with regard to metallic minerals, metals and alloys was around Rs 776 crores, out of which 80% was accounted for by ferrous materials. Though the need for increasing the value of our exports is very clear, a decision regarding the choice of minerals/metals which can be safely exported without impairing our future requirements, is not that straight-forward. It is obvious that we should export only those minerals whose resource position can be categorised as abundant. If one considers the metals in this category (Table 1), the export potential for sodium and magnesium is negligible because of their easy availability in the oceans of the world. Barytes and limestone, the minerals for metals barium and calcium, figure in the non-metallics section. The presently indicated resources of zinc-lead ores together with the recent discovery of a high grade (13 to 14%) zinc resource in the Rampura-Agucha area in Bhilwara district of Rajasthan, enable us to position zinc-lead in the abundant category; however, the production from these mines is yet to be organised. As regards radium, a radioactive metal, the absolute level of consumption is quite low and clear data are not available regarding its reserves position, production and consumption.

If we exclude the metals mentioned above, we still have as many as seven metals and their ores: gallium and aluminium (bauxite), titanium (ilmenite); chromium (chromite), iron (iron ore), zirconium (zircon), and light rare earths (rare earth oxides), for further consideration. To determine whether any of these minerals can be exported,

Table 5

Export of Metallic Minerals, Metals and Alloys

Mineral/metal/ Alloys	Quantity '000 Tonne	Value Rs Cr.	Percent share of total value
Iron Ore	30150.0	578.8	74.5
Iron & Steel	95.1	54.4	7.0
Chromite	229.7	38.2	4.9
Copper Ore & Conc.	45.4	26.6	3.4
Aluminium	13.5	24.8	3.2
Manganese Ore	486.4	20.4	2.6
Alumina	41.0	8.4	1.1
Chromium	8.6	7.3	0.9
Copper & Alloys	0.6	6.3	0.80
Brass & Bronze	1.1	3.2	0.40
Iron & Steel (SCRAP)	64.2	2.4	0.30
Ilmenite	18.0	1.8	0.20
Ferro Manganese	5.3	1.5	0.20
Others	---	2.0	0.25

Source: M.S. Division, Indian Bureau of Mines (year: 1985-86, value Rs 776 crores).

we need information on the proven reserves of the ore and its present level of production. The ratio of ore production to reserves can be taken as an index of the rate at which the resource is getting depleted.

Tables 6 to 11 contain data regarding the extent of proven reserves, the ore production (for the year 1985) and the computed ratio of production to reserves for bauxite, ilmenite, chromite, iron ore, zircon and rare earths respectively. In each of these tables, the data relevant to India are compared with those for countries which have substantial resources of the same mineral. In particular, countries which are leading exporters of the mineral are included. The computed ratio of production to reserve suggests that the present level of depletion of our reserves, with respect to iron ore, zircon and rare earths is comparable to the level of depletion of the reserves allowed by the

Table 6

Ratio of Production to Reserves in Leading Countries : Bauxite

Country	Proven Reserves [1] (Million Tons)* (1985)	Production [2] ('000 Tons) (1985)	Ratio of Production to Reserves $(\times 10^{-3})$
India	1,600	1,700	1.1
Guinea	5,600	14,740	2.6
Australia	4,440	32,180	7.3
Brazil	2,250	6,430	2.9
Jamaica	2,000	6,140	3.1
Guyana	700	2,130	3.0
Greece	600	2,360	3.9
Surinam	575	3,370	5.9
Yugoslavia	350	3,480	9.9
France	30	1,470	0.6

[1] *Minerals Facts and Problems*, 1985, U.S Bureau of Mines p.13.
[2] *Mining Annual Review* ,1986. p.44.
* Includes only measured and indicated.

other countries. In contrast, in the case of bauxite, ilmenite and chromite the rate at which we are utilising our reserves is significantly lower than the depletion rates observed in other countries. Thus, we are led to conclude that, in the case of bauxite, ilmenite and chromite, augmentation of our ore production is called for. Since the demand within our country for these ores cannot be expected to pick up dramatically in the coming years, the way to sustain the augmentation of ore production is through vigorous export. In fact such a scenario is already emerging in the case of bauxite. In the recent years, the augmentation of ore production, and the subsequent processing to alumina and aluminium by establishing NALCO in Orissa, has resulted in the export of both these value added products. In the year 1988-89, NALCO's export earning through export of alumina (over 3.8 lakh tonnes) and aluminium (about 15000 tonnes) has exceeded Rs.230 crores and covered leading world trading nations like Japan. Such a strategy needs to be encouraged and extended.

Table 7
Ratio of Production to Reserves in Leading Countries: Ilmenite

Country	1983 Reserves[1]		1985 Production[2] (1000 tons)		Ratio of Production to Reserves ($\times 10^{-3}$)	
	Ilmenite	Rutile	Ilmenite	Rutile	Ilmenite	Rutile
India	58[*]	5.2	55	4	0.9	0.8
South Africa (Slag)	25	2.4	214	35	8.6	14.6
Norway	21	-	158	-	7.5	-
China	20	-	45	-	2.3	-
Canada (Slag)	18	-	303	-	16.8	-
Australia	15	5.7	335	108	22.3	18.9
USA	7.9	0.2	W	W	-	W
USSR	4	1.6	144	6	36.0	3.7
Sri Lanka	2.5	0.5	29	5	11.6	10.0
Brazil	1.1	40.0	18	-	16.4	-

(Million short tons* of Ti contained). *Mineral Facts and Problems*, 1985, U S Bureau of Mines, p.860-864.

[1] Includes only measured and indicated.
[*] Assuming average content of 54% TiO_2 in Ilmenite.
w Withheld.

685

Table 8
Ratio of Production to Reserves in Leading Countries: Chromite

Country	Proven Reserves [1] (Million Tonnes)* (1983)	Production [2] (1000 Tonnes) (1985)	Ratio of Production to Reserves (x 10^{-3})
India	103	440	4.3
South Africa	913	3,340	3.7
USSR	142	2,560	18.0
Finland	19	506	26.6
Zimbabwe	19	480	25.3
Philippines	15	275	18.3
Brazil	9	310	34.4
Albania	7	920	131.4
Turkey	5	500	100.0
Iran	2	40	20.0

[1] *Minerals Facts and Problems*, 1985, U S Bureau of Mines, p.143.
[2] *Mining Annual Review*, 1986, p.69.
* Includes only measured and indicated.

Table 9
Ratio of Production to Reserves in Leading Countries: Iron Ore

Country	Proven [1] Reserves (Million Tons)*	1985 Production [2] (Million Tons)	Ratio of Production to Reserves (x 10^{-3})
India	12000	42.5	3.5
USSR	59000	248.0	4.2
USA	15800	48.8	3.1
Brazil	15600	120.0	7.7
Australia	15000	95.3	6.4
Canada	11700	39.0	3.3

Table 9 (*Contd.*)

Country	Proven [1] Reserves (Million Tons)*	1985 Production [2] (Million Tons)	Ratio of Production to Reserves (x 10^{-3})
China	9000	130.0	14.4
S.Africa	4000	24.4	6.1
Sweden	3000	20.6	6.9
France	2200	14.7	6.7
Venezuela	2000	14.8	7.4
Liberia	900	16.1	17.9

[1] *Minerals Facts and Problems*, 1985, U S Bureau of Mines, p.388.
[2] *Mining Annual Review*, 1986, p.55.
* Includes only measured and indicated.

Table 10

Ratio of Production to Reserves in Leading Countries: Zircon

Country	Reserves [1] in terms of Zirconium (million tons)*	1985 Production [2] (1000 tons) Zirconium [a]	Ratio of Production to Reserves (x 10^{-3})
India	6.2	6.5	1.0
South Africa	3.4	69.7	20.5
Australia	8.7	248.9	28.6

[1] *Mineral Facts and Problems*, 1985, U S Bureau of Mines, p.943.
[2] *Mining Annual Review*, 1986, p.89.
[a] Calculated from $ZrSiO_4$ (Zircon Sand).
* Includes only measured and indicated.

Table 11

Ratio of Production to Reserves in Leading Countries: Rare Earths

	Reserves[1] (Million Tons) of REO (1983)	Production ('000 +) of REO (1985)	Ratio of Production to Reserves (x 10^{-3})
India	2.20*	2.2	1.0
China	36.00	6.0	0.2
USA	4.90	17.1	3.5
USSR	0.45	1.5	3.3
Australia	0.20	8.0	40.0

[1] *Mineral Facts and Problems*, 1985, U S Bureau of Mines, p.650-652.
* Reserve comprises of mainly light rare earths.

Reckoning with the projected demands for the metals concerned during the next one hundred years, it appears safe to recommend the augmentation of export of bauxite by a factor of four to five, chromite by a factor of four to five, and ilmenite by a factor of 15 to 20. As illustrated in *Fig.3*, the augmentation of export of these minerals by the above factors will bring the static life time of our reserves to levels comparable to those prevalent in other countries also having substantial resources of these minerals. Although we have specifically mentioned three materials, namely bauxite, chromite and ilmenite, there are also clear possibilities for augmenting production and export in respect of iron ore, zircon and rare-earth based minerals.

The statement that the mineral resource is a valuable natural asset of any nation hardly needs to be made. Invariably, the attempt should be to export value added products. The market trend in leading trading nations is towards an upturn in metal prices. It is therefore just possible that, despite the higher indigenous cost of production, India can find export markets for converted products based on metallic materials. If a decision is made to export the surplus ores, careful thought has to be given, and an imaginative policy evolved, so that the export of such an asset brings back an equally precious return either

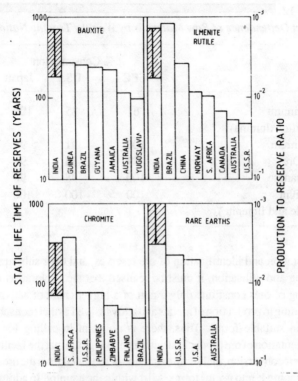

Fig.3 Effect of suggested augmentation of production on life-term of reserves (hatched).

in terms of a locally needed raw material or a strategic advantage for the country. In this context, it is relevant to note that major developed nations are largely import-dependent in regard to the same resources in which India is abundant (or nearly so) as illustrated in Table 12. This situation presents valuable opportunities. For instance, USA is today dependent on South Africa for its requirements of chromite (or chromium metal). India can take a decision to augment indigenous chromite production (as well as products based on this raw material) as a policy to become internationally competitive.

3   Elements of Policy for Abundant Metals

Though the minerals policy should apparently be concerned with the

Table 12

Import Dependence of Raw Materials by Wealthy Trading Nations

| | % Consumption | | |
	EEC	USA	Japan
Aluminium (Bauxite/Alumina)	61	85	100
Chromium	100	91	98
Iron Ore	79	36	99
Manganese	100	98	98
Titanium (Rutile and Ilmenite)	100	100	100

exploration and identification of ore reserves, and their subsequent mining and utilisation, it must be realised that the exploration and mining of ores constitute only a part of a bigger chain of activities extending to production of metals and alloys as well as their consumption in suitable forms. Thus, there is no point in calling for the augmentation of ore production without making sure that the facilities for their conversion, including extraction of the metal from the ore and for shaping it into useful forms, exist within the country. In addition, a proven demand for the metal or alloy must exist within or outside the country. Therefore, it is clear that before we formulate the elements of the national minerals policy, a comprehensive analysis needs to be carried out for each metal with regard to the various segments in the activity chain, from exploration of the ore to the production of metal and its application. The policy should then aim at removing the bottlenecks so that the indigenous picture for the metal appears smooth from whatever angle the same is viewed. In this section, three metals, namely iron, titanium and light rare earths will be considered to highlight the specific features requiring attention in each case. The exercise will naturally bring out considerations that should underlie a national minerals policy. It is interesting to point out that these three metals, in which India is rich, have received varying degrees of attention. Iron and steel represent a metal for which the mining and extraction facilities are well developed but still we import

Table 13
Import of Metallic Minerals, Metals and Alloys

Mineral/Metal/ Alloys	Quantity '000 Tonne	Value Rs Crores	Per cent Share of Total Value
Iron & Steel	2302.1	1308.3	58.4
Iron & Steel (Scrap)	1331.7	258.2	11.5
Copper &Alloys	110.6	225.3	10.0
Aluminium	70.0	109.4	4.9
Ferro-Nickel	23.3	54.9	2.4
Nickel & Alloys	6.4	41.3	1.8
Tin & Alloys	2.0	27.0	1.2
Non-Ferrous Meta l Scrap	30.2	25.7	1.1
Lead & Alloys	41.7	24.4	1.1
Copper (Scrap)	15.2	23.0	1.0
Ferro-Chrome	11.4	21.5	1.0
Brass & Bronze (Scrap)	13.8	15.9	0.7
Ferro-Manganese	1.0	15.4	0.7
Brass & Bronze	2.8	10.4	0.5
Titanium Dioxide	5.3	9.8	0.4
Nickel Ores & Conc.	3.4	8.2	0.4
Tin	12.3	6.9	0.3
Pig & Cast Iron	21.4	5.2	0.2
Zinc Ores & Conc.	0.90	5.0	0.2
Vanadium Ores & Conc.	0.90	4.5	0.2
Tungsten Alloys	0.05	4.2	0.2
Others	-	37.5	1.8

(Year: 1985-86, Value Rs 2242 Crores)

in one form or another a significant fraction of our consumption. In fact, as illustrated in Table 13, about 70 per cent of our annual import bill of Rs 2242 crores for the year 1985-86 in the case of metallic minerals, metals and alloys was accounted for by iron and steel imports. Titanium represents an intermediate case where the setting up of mining and beneficiation facilities has been initiated but no full-scale metal production facility still exists. However, the light rare earths represent the other extreme where we have not yet achieved individual metal production on any reasonable scale.

To facilitate the analysis indicated above, a flow chart has been prepared for each metal depicting the various activities and their mutual interaction. Each box represents one activity, and related activities are joined by "arrowed" lines. There are accompanying indications to show if an activity has already been developed to a satisfactory level in the country, is yet to be fully developed or is one demanding urgent attention. *Figures 4, 5, and 6* represent such flow charts for three abundant metals, namely iron and steel, titanium and light rare earths which are self-explanatory. Iron and steel will not be

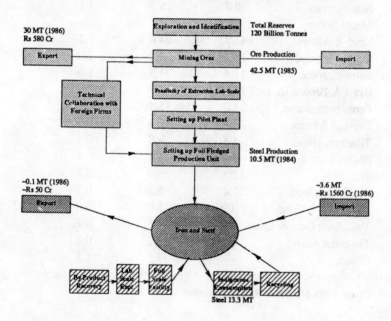

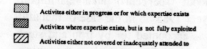

Fig.4 Flow chart for iron and steel.

discussed further as a special task force has been set up by the Department of Steel, Government of India, to discuss and prepare a report on national missions for this area. Titanium and rare earths are discussed further as examples illustrative of actions needed to deal with the ore-to-product chain in a comprehensive manner.

3.1 Titanium

We shall discuss the case of titanium metal at some length. Despite the fact that India is richly endowed with vast resources (especially ilmenite in the coastal belts) of this light engineering metal with exceptional physical and chemical characteristics, titanium does not seem to have received adequate attention in the country. This

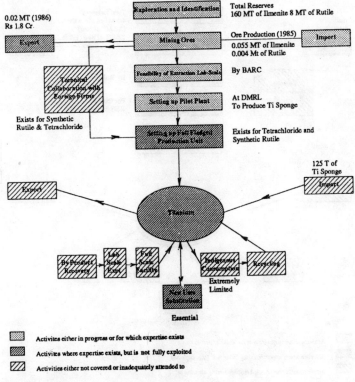

Fig.5 Flow chart for titanium.

situation needs to be corrected. Titanium is a jewel of a metal which India is fortunate to possess in abundance. This non-ferrous metal should prominently figure in the indigenous minerals policy. In the formulation of an integrated minerals policy, extending from ore to product, the potential for titanium is immense both for applications appropriate to its own characteristics as well as in areas where titanium can be a substitute for metals not available in our country such as nickel.

The current titanium scene in India is depicted in *Fig.7*. The following salient features emerge from this chart.

● A well established beach sand industry for mineral separation exists in the country.

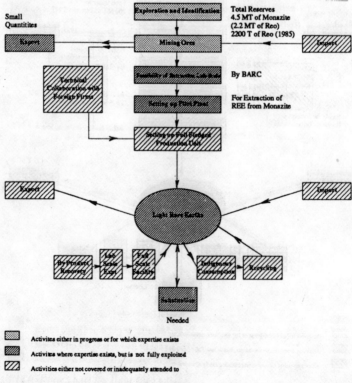

Fig.6 Flow chart for light rare earths.

- Large scale production facilities for synthetic rutile, titanium tetrachloride and pigment grade titanium dioxide are already in operation.

- No commercial plant exists for the production of titanium sponge metal and the country's requirement is met entirely by imports.

- Capacity for mill products production exists at MIDHANI.

- Technology for equipment fabrication is available but needs augmentation in respect of power plant condensers.

The flow chart of activities in respect of titanium is presented in Fig.5. As mentioned earlier, the first and foremost task is to establish a full-

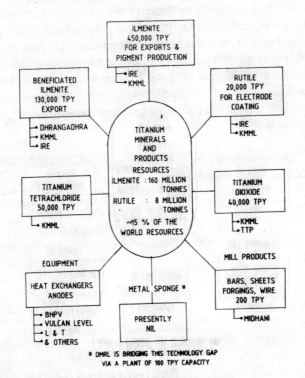

Fig.7 The Indian titanium scene (1986).

fledged production facility for titanium sponge metal with a capacity of say 1000 TPY initially, for meeting the indigenous requirement. While there is scope for increased production of the minerals for export, as recommended in an earlier section, a serious effort is needed for converting our mineral wealth into value added products such as pigment grade titanium dioxide and titanium metal for domestic use and export. Till now, technology for metal production on an appropriate scale was not available for ready exploitation. However, this gap has now been bridged with the establishment of the technology development centre at the Defence Metallurgical Research Laboratory, Hyderabad. Thus the country is now ready for establishing a full scale production plant based entirely on indigenous technology.

At present the extent to which titanium is used in our country is insignificant when compared to the levels of consumption reported by the developed countries. In Table 14 the world titanium sponge capacity is presented. The fact that   Japan, which does not have any mineral resources for titanium, is the second largest producer of this metal is truly astonishing. In Table 15 the projections of titanium requirements by the year 1990 and 2000 are brought out. As far as titanium is concerned, it is necessary to adopt a deliberate policy of nurturing its growth and promoting its widespread use by taking the following steps:

Table 14

World Titanium Sponge Production Capacity

Country	Annual Capacity (In Metric Tonnes)
USSR	45,000
Japan	35,000
USA	32,000
UK	5,000
China	3,000
Total	1,20,000

Table 15

Demand Forecast of Titanium (All Figures in Metric Tonnes)

	1983	1990	2000
USA	7,300	23,700	41,000
Rest of World	52,700	81,000	1,45,000
Total	60,000	1,04,700	1,86,000

● Increase user awareness of the exceptional properties of titanium: high specific strength, high specific toughness and excellent corrosion resistance.

● Encourage its use, in place of stainless steels, in chemical plant equipment and piping systems (this will substantially bring down import of nickel for stainless steel manufacture).

● Adopt a policy of using titanium tubing for all power plant condenser systems.

The demand for titanium mill products by the year 1989-90 is estimated at 715 tonnes, as per details in Table 16, which requires around 1400 tonnes of primary metal. This projection assumes a limited use of only around 200 tonnes of titanium tubing by the power generation industry. In this context it is to be noted that one of the main industrial uses of titanium is in the form of welded titanium tubing for steam condenser applications in power plants for reasons recorded in Table 17. From Table 18 it can be seen that more than 100,000 MW(e) of power has been generated using titanium tubed condensers based on a detailed techno-economic evaluation. It is to be emphasised that condenser tube failure accounts for as much as 38 per cent of all power plant outages. The experience abroad, particularly in the UK, Japan, France and the USA, has clearly indicated that steam condensers fabricated with titanium tubes, even in coast-based power plants, can operate without shut-down during the entire design life of a power station, i.e., about 40 years.

Table 16

Expected Annual Demand for Titanium by 1989-90

Sector	Expected Annual Demand by 1989-90 (Tonnes of Finished Products)
Aeronautics	80
Space	15
Power Generation	200
Petroleum and Petro-chemical	50
Electroplating	100
Heat Exchangers	50
Miscellaneous	20
Total	715

Titanium sponge requirement for above 1,430 TPY.

Table 17

Titanium Tubed Condensers for Power Generation Industry

1. Excellent corrosion resistance to saline waters.
2. Outstanding erosion resistance permitting higher water velocities.
3. Acceptable thermal performance over prolonged periods.
4. No retubing necessary during life-time of power plants. (~40 years) (No shut down on this account).
5. Zero leak rate - essential for nuclear power plants.
6. Substantial cost savings over the power station life-time.

In Table 19 a cost comparison of alternate condenser tube materials for 500 MW(e) sea water cooled power plants is presented, taking into account the current FOB prices for these materials. It is clearly seen from this table that titanium is the obvious choice particularly when the "cost of power not produced" due to outages for plugging/retubing is taken into account. If the Government gives a clear directive on the usage of titanium tubes for power plant condensers, the following advantages arise:

Table 18

Titanium Condensers in Power Industry

Country	Power - MW (e)
United Kingdom	26,000
Japan	21,000
France	15,500
Other Non-Communist Countries	37,000
India	500 *
Total	100,000

Titanium tubing installed 16,000 Tonnes

* Tata thermal power station, Bombay.

Table 19

Cost Comparison of Condenser Tube Materials in Sea Water Cooled 500 MW (e) Power Plant

	Titanium	Material[1] Cupro-Nickel	Al-Brass
Density gm/CC	4.5	8.9	8.4
Length of tubing (Km)	38.0	38.0	38.0
Size of tube (OD x WT in mm)	25.4 x 0.7	25.4 x 1.6	25.4 x 1.6
Cost of tubing/Metre (Rs)	63	85	58
Initial cost of tubing For one plant (Rs Crores)[2]	2.4	3.2	2.2
Number of replacements during life-span of 40 years	Nil	Two(Min.)	Three (Min.)
Cost of replacement -Rs Cr	Nil	6.4	6.6
Total Cost (Rs Cr.)	2.4	9.6[3]	8.8[3]

[1] Stainless steel unsuitable for sea water application.

[2] Does not include cost of other condenser components and fabrication.

[3] Does not include cost of power not produced due to outages for plugging/ retubing which is *very substantial*.

- Recurring imports of condenser tubes and condenser tube materials is totally avoided.

- There will be a substantial improvement in power station availability due to absence of outages for plugging/retubing.

- Substantial savings in long term power generation costs assured.

- Self-reliance in the manufacture and application of titanium based products.

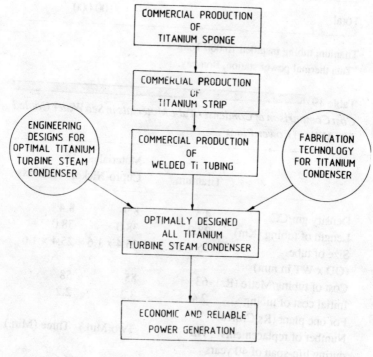

Fig.8 *Titanium condenser for economic power.*

In *Fig. 8* the constraints for large scale manufacture of titanium tubed steam condensers are projected. It can be seen that engineering designs of such condensers and the augmentation of fabrication technology through imports or indigenous developmental effort is essential.

While emphasis has been laid on the use of titanium tubes for power plant condensers, the potential for its widespread application is large and ever-expanding. Traditionally titanium has been chosen for aerospace applications. In this context it is interesting to point out that, in the USA, titanium sponge is therefore included in the list of critical materials for stockpiling. The metal is requisitioned by the US Government through defence production act contracts. It is reported that the US stockpiling of titanium in 1984 was as much as 33,150 tonnes while the target set by the Federal Emergency Management Agency (FEMA), based on a 3-year war criterion, for establishing stock pile objectives is a phenomenal 1,95,000 tonnes.

Indigenous capability to produce aerospace components correspondingly assumes considerable significance due to their strategic nature. The chloralkali industry, the petro-chemical industry, the fertilizer industry and the paper and pulp industry use a variety of equipment fabricated in titanium. For a country which has to import every gram of its nickel requirement, the indigenous availability of titanium should be reckoned as balancing.

Titanium's usage is not restricted to just the few industries mentioned above. The Soviets have successfully exploited its combination of properties for building deep diving ocean going vessels such as submarines, utilising several thousand tonnes of titanium in the process.

The Japanese, in particular, have come out with consumer products such as ball pens, table ware and bicycles in titanium that will last a life time and more.

Bio-compatability of titanium has led to its widespread usage for prosthetic applications including such vital parts as heart valves.

For a country like ours, that depends entirely on imported paper for printing currency, titanium coinage is an alluring alternative. With its light weight, high strength, excellent corrosion and wear resistance, titanium is an ideal coinage metal and requires serious consideration not only for its own merits but also as a substitute for scarce nickel whose import prices are constantly on the rise.

In conclusion, the following steps are recommended:

- Production plants to be set up immediately for titanium sponge and welded titanium tubing, for meeting domestic demands and exports.

- Once titanium tube products become available, a policy decision must be taken to use titanium tubed condensers in all future power plants, both sea-water cooled and river-water cooled.

- The latest technology for the design and fabrication of titanium steam condensers to be developed.

- An imaginative stockpiling policy for titanium sponge, similar to the one in operation in the USA, to be evolved for nurturing the growth and protecting the titanium sponge industry at least during the formative years.

Interestingly, the total investment required to place   titanium on a firm footing is less than Rs 100 crores for bridging all the gaps in the ore to product picture. This amount can cater for

- 1000 TPY titanium sponge plant
- a welded titanium tube manufacturing facility and
- augmentation of facilities for power steam condenser design and manufacture.

With such a not-too-large investment, the titanium metal scene will be rendered complete. Substantial revenue benefits are also antici-pated just from the power sector.

3.2 Rare Earths

The example of rare earths in the present discussion, as already mentioned, constitutes an extreme instance of a national resource which has remained largely untapped for metal production (see Fig. 6).

The country is endowed with sizeable natural resources of rare earths in the beach sand deposits of the south west and eastern coasts. The Indian Rare Earths, one of the major   producers of monazite in the world, would be reaching an annual production of 8000 tons when the Orissa plant (OSCOM) goes into full production.

The problem with the rare earth metals is their close similarity in chemical properties and the resultant difficulty in separating individual elements with a high degree of purity. The onset of the nuclear age has given much fillip, elsewhere in the world, to the development of rapid and effective methods of separation. Techniques like ion-exchange chromatography and solvent extraction, originally intended for separating trace level rare earth fission products from the fission of uranium, are today precisely the methods adopted for large scale production of individual rare earths.

India has established the capability in separation techniques. However, no great effort is under way, even at the present time, for undertaking any substantial scale of production.

The metallurgy of rare earths for extraction and refining of individual and mixed rare earths needs to be vigorously built up, if the present level of indigenous consumption of rare earth minerals, which is less than 10 per cent of the production, is to be meaningfully raised.

As in the case of titanium, the potential for industrial applications of rare earths, e.g. in electronics and optical industry, as catalysts in the petro-chemical industry and for pollution control have not been widely appreciated in India. There are further application areas. Rare earth additions to a variety of non-ferrous alloys are becoming increasingly popular for enhancing oxidation resistance and for grain refinement. In the new emerging class of rapidly solidified aluminium alloys (like Al-Fe-Ce alloys), rare earth additions are proving to be effective in accentuating undercooling effects and globularising precipitate morphologies. In the important area of permanent magnetic materials, neodymium-iron-boron magnets (see K.H.J. Buschow, *Materials Science Reports*, Vol.1, September 1986), based on the ternary intermetallic compound $Nd_2Fe_{14}B$, have generated considerbale technological interest. India should move expeditiously into the development of this and other rare-earth based products. These should not remain a virgin field in a land rich in natural resources as well as capable of mastering the necessary technologies. Again the minerals policy should encompass the gamut of issues pertaining to the ore at the one end and the usable product at the other.

4   Element of Policy for Strategic Minerals

The minerals policy of any country should identify certain materials as being of strategic importance and develop an exclusive policy for them. Clearly, the first step is the *identification* of these strategic materials. A U.S. source (OTA Report No. OTA-ITE-248, May 1985, U.S. Government Printing Office, Washington D.C) defines strategic materials as *those for which the quantity required for essential civilian and military uses exceeds the reasonably secure domestic and foreign supplies, and for which acceptable substitutes are not available within a reasonable period of time.*

We shall accept this definition in view of its general applicability. Although quite a few metals in the medium and insufficient resource categories (Table 1) qualify for consideration as strategic materials, the present discussion will be limited to five metals, namely

● Tungsten
● Platinum group of metals
● Lithium
● Cobalt and
● Nickel

Of the above, tungsten should occupy the most prominent place as a strategic metal for reasons which we will explain later.

The above choice has been made so as to bring out distinctly the different aspects that call for concentrated attention. For instance, in the case of tungsten, the platinum group and lithium, the emphasis has to be in the field of exploration and mining since the country is yet to be assured of adequate resources in these cases. For cobalt, the problem is to generate metal production from by-products resulting from the present copper and zinc operations. On the other hand, in the case of nickel, for which substantial resources are available, we have no technology yet, for extraction from the available lean ores.

Since the first three metals listed above require intensification of efforts in the exploration and mining areas, the next section discusses a general strategy for exploration for our country. This will be followed by a brief consideration of the necessary activities related to each of the above five metals.

4.1   Strategy for Exploration

The mineral exploration activity in our country will be greatly facilitated through the adoption of modern theoretical approaches, involving modelling methodologies, better data generation and dissemination and regular and widespread application of current exploration technologies. Thus we need to develop a strategy for exploration itself which should have the following features:

● Choice of favourable target areas on the basis of relevant geological modelling.

● Speedy publication of geological maps.

● Generation of regional geophysical and geochemical data, especially gravity maps and magnetic maps, so that geological models become more meaningful and relevant.

● Development of polymetallic deposits for extraction of the major metals, and the low content metals aggregating to economically workable levels.

We shall cite some relevant examples:

O   U, Cu, Ni and Mo lodes occur together in Eastern Bihar.
O   Au and W occur together in Kolar.

(Past mining practices did not lay sufficient emphasis on scheelite recoveries. The situation has now been corrected).

● Application of modern exploration technologies like

O   Introduction of rapid in-field analytical techniques such as portable XRF to speed up generation of geochemical data.

O   Computer processing of the large geological and geophysical data already available and endowing geophysical instruments with greater capabilities for storage and interpretation of data.

O   Improvement in drilling and bore-hole logging technology to generate a larger database. These essential features pertaining to scientific inputs in exploration are brought out in *Fig. 9*.

4.2   Tungsten

We have a situation in regard to tungsten which may be considered as diametrically opposite to that pertaining to titanium (section 3 of this report). In the case of titanium, the resources have been identified and are plentiful. The production of the mineral on a large scale is in progress. However, the country is yet to take advantage of this abundant resource position for product development and utilisation in critical areas, such as the power generation sector. In contrast, we have gone much ahead in regard to the development of products and components made of tungsten. The Ministry of Defence is presently setting up a major production unit based on tungsten. There are other industries already functioning which use tungsten-based materials such as the alloy steel and cutting tool industries. However, the

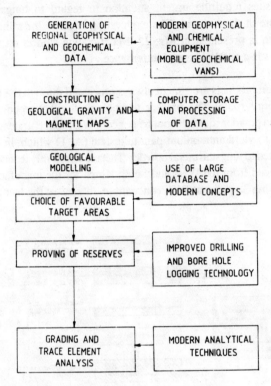

Fig.9 Strategy for exploration based on scientific inputs.

tungsten mining activity in the country is in a stage of infancy (*Fig. 10*). Deposits of tungsten are indicated in various parts of the country (*Fig. 11*). However, the only mine operating currently is the one at Degana in Rajasthan producing about 50 tonnes of wolframite per annum. The beneficiation methods practised at Degana involve antiquated practices such as hand panning. It is only recently that a pilot beneficiation plant has been installed here and much more needs to be done to augment the output of the mine as well as evaluate the viability of exploiting the substantial resources indicated in the granite rock of the region. Bharat Gold Mines Limited have also set up a pilot plant for beneficiation of their scheelite. Beneficiation studies on the other ore deposits are still to be carried out.

The issues requiring urgent attention in regard to tungsten are therefore at the ore end. The exploration strategy for minerals of tungsten is presented in *Fig. 12*. Beneficiation studies on various deposits is a parallel area of importance.

Tungsten products are almost always processed through the powder metallurgy route. The present technology for the production of tungsten powder rquires about 65 per cent WO_3 concentrate which is treated to yield ammonium para tungstate (APT) which, in turn, is converted to tungsten metal powder. There are presently technologies announced from the West and South Korea for the manufacture of APT using 30 per cent WO_3 concentrate. In view of the fact that the

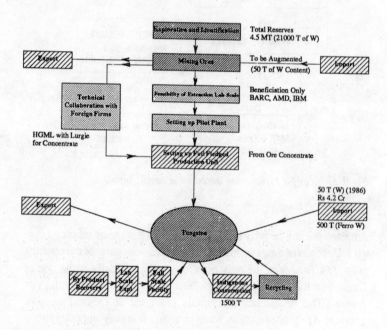

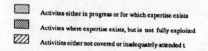

Fig.10 Flow chart for tungsten.

indigenous tungsten deposits contain only 0.1 to 0.3 per cent WO_3, such new technologies for the production of APT need to be looked at and adopted.

Tungsten is a metal which is amenable to recovery from scrap. In the United States as much as 30 per cent of the tungsten requirement is met through scrap recoveries. India can ill-afford not pursuing such means of generating tungsten metal.

In view of tungsten being a strategic product, and its as-yet underde-veloped resource position, there is a strong case here for stockpiling the raw materials needed through imports. The present time seems to

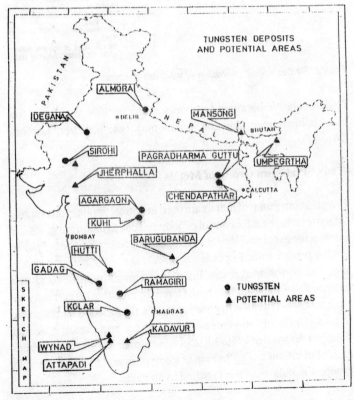

Fig.11 Tungsten deposits and potential areas.

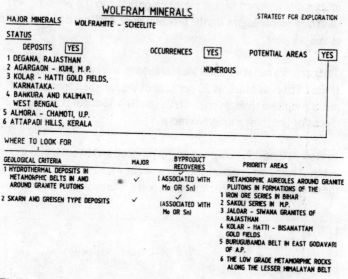

Fig.12 *Strategy for exploration of wolfram minerals.*

be particularly opportune for a stock piling decision in view of low international price situation which may not, however, last long (*Fig. 13*).

4.3   Platinum Group of Metals

The platinum group of metals are used in critical applications such as catalytic converters for control of pollution due to industrial gases and automotive exhaust emissions. They are irreplaceable as catalysts in certain petrochemical operations. In addition they are used extensively as corrosion resistant liners in chemical industries and in a variety of electrical applications. Given the facts that (a) currently we have no resources of the Pt group of metals, (b) these are irreplaceable in important application areas and (c) all our requirements are met through imports (for 1989-90, import is estimated at 9000 troy ounces each of platinum and palladium at a cost of over Rs 11 crores) the Pt group of metals must be classified as strategic.

As on date we do not have any clearly identified deposit of the

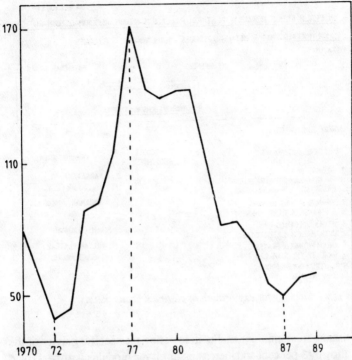

Fig.13 Price history of tungsten concentrate ($ /MTU of concentrate). 72-77 Steep price hike, 77-87 Extreme downfall, 87-89 Anticipated uptrend in price. (Source: NML, Jamshedpur).

platinum group of metals in the country. Hence the priority should be the identification of potential areas followed by a detailed exploration in those areas. Fig. 14 indicates the geological criteria to be employed while selecting areas. The same information for platinum only is also presented in Fig. 15 wherein the various potential areas are marked on the India map.

4.4   Lithium

Lithium is a metal of the future. With a density as low as 0.53 g/cc, this lightest metallic element has been recently discovered to be a wonder alloying addition to aluminium. Every unit addition of

711

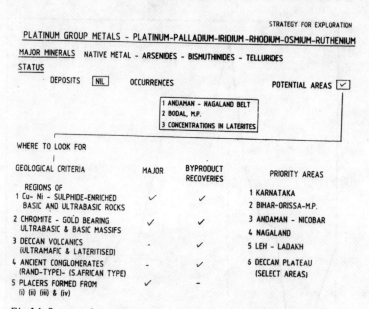

STRATEGY FOR EXPLORATION

PLATINUM GROUP METALS - PLATINUM-PALLADIUM-IRIDIUM-RHODIUM-OSMIUM-RUTHENIUM

MAJOR MINERALS NATIVE METAL - ARSENIDES - BISMUTHINIDES - TELLURIDES

STATUS

DEPOSITS [NIL] OCCURRENCES   POTENTIAL AREAS [✓]

1 ANDAMAN - NAGALAND BELT
2 BODAL, M.P.
3 CONCENTRATIONS IN LATERITES

WHERE TO LOOK FOR

GEOLOGICAL CRITERIA	MAJOR	BYPRODUCT RECOVERIES	PRIORITY AREAS
REGIONS OF			
1 Cu- Ni - SULPHIDE-ENRICHED BASIC AND ULTRABASIC ROCKS	✓	✓	1 KARNATAKA
2 CHROMITE - GOLD BEARING ULTRABASIC & BASIC MASSIFS	✓	✓	2 BIHAR-ORISSA-M.P.
3 DECCAN VOLCANICS (ULTRAMAFIC & LATERITISED)	-	✓	3 ANDAMAN - NICOBAR
4 ANCIENT CONGLOMERATES (RAND-TYPE)- (S.AFRICAN TYPE)	-	✓	4 NAGALAND
5 PLACERS FORMED FROM (i) (ii) (iii) & (iv)	✓	-	5 LEH - LADAKH
			6 DECCAN PLATEAU (SELECT AREAS)

Fig.14 Strategy for exploration of platinum group metals.

lithium to aluminium results in lowering the specific gravity of the alloy by 3 per cent and increasing its elastic modulus (stiffness) by 6 per cent. High strength aluminium-lithium alloys are thus the latest to enter the aircraft industry. Large scale commercial production is at an early stage even in the world's technologically leading countries. Another major use of lithium in the aluminium industry is the addition in the form of lithium carbonate to the cryolite bath in aluminium potlines.

Lithium is a very active electrochemical metal which provides more energy in a smaller space as anode in batteries than all other rival materials.

These and other applications of lithium mark out this metal as one of immense futuristic potential.

India should take a lead in this strategic metal. The extraction of lithium through the fused salt electrolysis process has already been

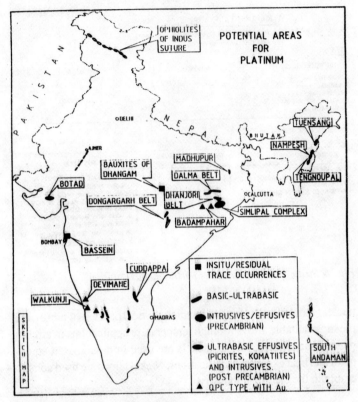

Fig.15 Potential areas for platinum.

attempted by Bhabha Atomic Research Centre. Larger scale adoption of this process has been taken up by the Defence Metallurgical Research Laboratory.

The paramount need is in the area of exploration and identification of the resources for this important metal. A strategy for exploration is suggested in *Figs. 16 and 17.*

4.5 Nickel

Nickel is at present an important constituent of austenitic stainless steel which is used extensively by the transportation, chemical and

713

ALKALI METALS - Li - Cs - Rb

STRATEGY FOR EXPLORATION

MAJOR MINERALS

Li = AMBLYGONITE - SPODUMENE - LEPIDOLITE - PETALITE
Cs = POLLUSITE - Rb = NO MINERAL

STATUS

DEPOSITS ☑   OCCURRENCES ☑   POTENTIAL AREAS ☑

1 BIHAR MICA BELT
2 MUNDWAL, M.P.   FEW
  KORAPUT, ORISSA
3 KADAVAL (RATNAGIRI)
  MAHARASHTRA

WHERE TO LOOK FOR

GEOLOGICAL CRITERIA	MAJOR	BYPRODUCT RECOVERIES	PRIORITY AREAS
1 SUB-SURFACE BRINES	✓	-	NO SUB-SURFACE BRINE DEPOSITS WITH Li KNOWN IN INDIA -
2 LATE SODIC PHASES OF GRANITE PEGMATITES	✓	-	1 SALINE DEPOSITS OF MANDI, H.P AND ALL THE SALINE LAKES MAY BE INVESTIGATED
3 GREISENISED ZONES IN GRANITES	-	✓	2 AREAS OF MARINE SEDIMENTS IN GUJARAT WITH WAPORITE SEQUENCES
			3 MAJOR PEGMATITE BELTS OF INDIA (IN RAJASTHAN, BIHAR, ORISSA, M.P., A.P. KARNATAKA AND TAMIL NADU
			4 Cs AND Rb CONTENT IN ALKALI BERYLS FROM RAJASTHAN AND BIHAR

Fig. 16 Strategy for exploration of alkali metals.

nuclear industries. Superalloys, which have 50 per cent of nickel, are the most suitable alloys for high temperature applications in aircraft gas turbine engines. Cupro-nickels are used in desalination equipment and heat exchanger applications. Nickel alloys are used considerably in the electrical industry.

The fact that we have resources predominantly of low grade lateritic ore, when added to the irrefutable importance of nickel to the economy, renders nickel a strategic metal.

The Sukinda (in Orissa) laterite has a nickel content in the range 0.2 to 1.2 per cent, which is much smaller than the nickel levels found in deposits in other countries. Further, beneficiation efforts with the Sukinda ore have so far been unsuccessful because this deposit is characterised by dissimilar suites of nickel having widely varying minerological and chemical characteristics. A silver lining in the otherwise dark picture for nickel is a sizeable amount of nickel (about 0.5 per cent) present in overburdens generated during chromite mining in the Sukinda belt itself. These chromite overburdens seem to be capable of generating about 15000 tonnes of nickel every year.

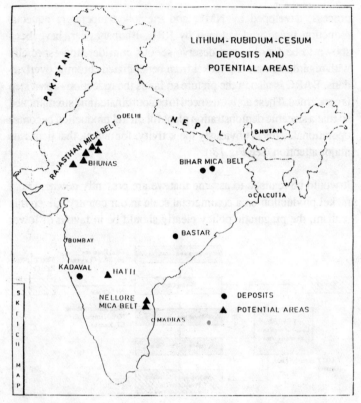

Fig.17 Lithium-Rubidium-Cesium deposits and potential areas.

This proposition looks attractive since mining costs will be small.

RRL, Bhubaneswar, has developed a process by which the chromite overburdens can be concentrated to yield an ore containing 1.1 per cent Ni consistently.

As regards the extraction of nickel from the above ores, a commercial plant in collaboration with Elken Laboratory, Norway is being set up in Orissa to extract ferro nickel from Sukinda nickel ore using pyrometallurgical processes.

Two other processes, namely the reduction roast - ammonia leach process, developed by NML, and the low temperature aqueous reduction process, developed by RRL, Bhubaneswar, have been shown to be feasible and deserve serious consideration especially with regard to extracting nickel from beneficiated chromite overburdens. BARC is also in the picture so far as the reduction - roast step is concerned. These activities need to be coordinated and strengthened so that a sizeable demonstration plant for nickel production becomes operational. This is obviously the activity, for nickel, that demands major attention (see *Fig.18*).

It would be realistic to assume that we are presently nowhere near nickel production on a commercial scale in our country. Given this realism, the pragmatic policy clearly should be in favour of lower

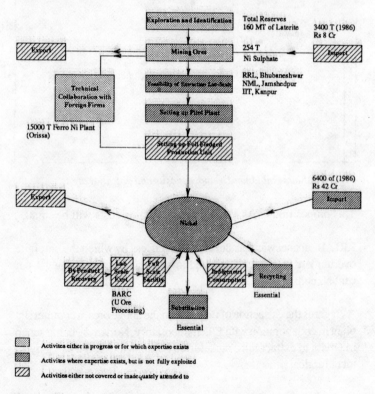

Fig.18 Flow chart for nickel.

consumption of this metal, and its alloys, through substitution wherever possible. The minerals policy for nickel then should actually direct its emphasis at the product end to ensure that less of the metal is consumed and imports are lowered. Low nickel and nickel-free austenitic stainless steels in place of high nickel austenitic stainless steels, titanium in place of stainless steel and cupronickels, development and use of nickel-free iron-based alloys (eg. Kanthal) for heating element applications and Fe-Nd-B alloys for permanent magnets, and the elimination of nickel from coinage should be the type of moves that can be considered for decisions of policy.

4.6   Cobalt

Production of cobalt, as the main metal, is dependent on a relatively high grade ore body. No such ore bodies have been discovered in our country. The promising source for cobalt is in the chromite overburdens. An estimate, with reference to the current rate of chromite ore production in the Sukinda belt, indicates a possible production of 1500 tons of cobalt as a byproduct every year. This possibility is yet to be demonstrated.

Hindustan Zinc Limited (HZL) is generating about 300 tonnes annually of what is known as beta cake at their Visakhapatnam plant and 200 tonnes of the same at their Debari plant. The average cobalt content in this residue is 1 to 2 per cent. The stockpile of the cake is over 2500 tonnes. HZL, in collaboration with BARC, is engaged in developing a solvent extraction process to produce cobalt sulphate or cobalt oxide. The metal can be obtained subsequently through an electrowinning operation.

Indian Copper Complex (Ghatsila) generates about 50 tonnes per day of copper converter slag containing 0.5 to 1 per cent cobalt and 0.5 - 2.6 per cent nickel. Hindustan Copper have been working on a process involving slag flotation, digestion, leaching, purification, concentration and separation of cobalt and nickel. Regional Research Laboratory, Bhubaneswar has been working on the extraction process for cobalt and nickel from the copper converter slag, utilising solvent extraction and pressure leaching.

717

The case of cobalt thus highlights the area of byproduct recovery (*Fig. 19*) of metal values from the existing operations for other metals such as copper and zinc. This feature contributes another aspect to the formulation of a comprehensive policy. Consumption of this scarce metal is amenable to reduction. New magnetic materals such as Nd-Fe-B do not contain Co. The hot isostatic process is capable of consolidating carbide products without resorting to liquid phase sintering thus obviating the presence of Cobalt.

5   Status of Non-metallic Minerals

The non-metallic minerals are not discussed in detail to avaoid a repetition of statements and suggestions already made in the context

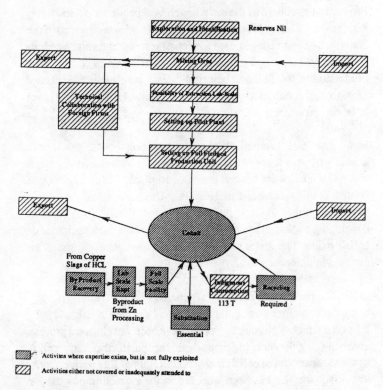

Fig.19 Flow chart for cobalt.

of metallic minerals. However, essential data are presented in Tables 20 to 23.

A classification of non-metallic minerals as (a) abundant, (b) medium and (c) insufficient is shown in Table 20. The index of depletion, used earlier, defined as the ratio of production to reserves for the non-metallic minerals is given in Table 21.

Non-metallic minerals are relatively low-value materials. The relevant data regarding our export of non-metallic minerals during the year 1985-86 is given in Table 22. The principal minerals being

Table 20

Status of Indian Non-Metalic Mineral Resources

Abundant	Medium	Insufficient
Magnesite	Ochre	Kyanite
Fireclay	Fluorspar	Sillimanite (Lumpy)
Sillimanite (Granular)	Sateatite	Bauxite (Refractory)
Dolomite	Clcite	Chromite (Refractory)
Zircon	Diamond	Graphite
Kaolin		Felspar
Limestone		Gypsum
Corundum		Asbestos
Garnet		Rock Phosphate
		& Apatite
Mica		Potash
Barytes		Vermicullite
Bentonite		
Wollastonite		
Silica		
Pyrophyllite		
Dimension Stones		

* Source: *National Minerals Inventory*, 1985, prepared by Indian Bureau of Mines.

Table 21
Reserves of Non-metallic Minerals (as on 1985-86)

Minerals	Reserves (Million Tonnes)	Ore Production ('000 Tonnes)	Ratio of Production to Reserves (10^{-3})
Limestone	70,000	53.0	0.001
Mica	-	32.5	-
Dolomite	4,608	2197.0	0.50
Gypsum	319	1722.0	5.40
Kaolin	872	747.0	0.85
Fireclay	703	701.0	1.00
Magneste	222	460.0	2.10
Apatite & Rock Phosphate	117	656.0	5.60
Barytes	71	337.0	4.75
Calcite	2.5	31.0	12.40
Kyanite & Sillimanite	36	51.0	1.40
Talcisteatite	76	379.0	5.00
Vermicullite	0.3	4.0	13.30
Graphite	4.7	38.0	8.10
Asbestos	1.5	27.0	18.0
Fluorite	12.0	22.0	1.83
Diamond	1.4 (Million Carats)	6.0 (Million Carat)	4.30
Pyrite	85	74.0	0.90
Potash	16	-	-
Silica Minerals	780	1339.0	1.70
Felspar	15	48.0	3.20
Wollastonite	2.4	24.0	10.0
Ochre	5.9	85.0	14.4
Flouspar	1.9	20.0	10.5
Pyro Phyllite	4.3	66.0	15.3
Corundum	0.14	0.60	4.3
Garnet	46	5.4	0.12
Bentonite	276	105.0	0.38
Fullers Earth	236	85.0	0.36

Source: National Minerals Inventory, 1985, Indian Bureau of Mines.

Table 22

Export of Non-metallic Minerals

Mineral	Quantity ('000 tons)	Value (Rs in Crores)	Per cent Share Total Value
Mica (all types)	67.8	72.3	45.2
Building and Monumental stones	358.0	53.7	33.6
Barytes	375.0	14.2	8.9
Quartz	70.8	4.0	2.5
Emerald	-	2.0	1.2
Slate	7.4	1.8	1.1
Bentonite	21.4	1.6	1.0
Steatite	12.4	1.5	0.9
Limestone	230.5	1.0	0.6
Others	-	8.0	5.0

(Year 1985-86, Value Rs 160 Crores). Source: M.S. Division, Indian Bureau of Mines.

exported are mica, barytes, quartz, limestone and building materials. The imports were valued at Rs 481 crores during the year 1985-86 with sulphur and rock phosphate dominating and accounting for 78 per cent of the import value (Table 23).

The approach to exploitation of non-metallic minerals, to derive greater benefit for the national economy is governed by the same basic principles as those underlying the metallic ones, some of which have been discussed in this report. Tables 20-23 can help in analysing the present situation and guiding the future strategies.

Again, rock phosphates present an excellent illustration of what needs to be done. The analysis, based on the ore-to-product philosophy, clearly suggests (a) further exploration to discover higher grades (e.g. > 31 per cent P_2O_5, <10 per cent SiO_2, and < 1 per cent MgO) of phosphate; (b) development of beneficiation technologies and even

721

Table 23
Import of Non-metallic Minerals

Mineral	Quantity ('000 tons)	Value (Rs in Crores)	Per cent Share Total Value
Sulphur	989.9	220.3	45.8
Rock Phosphate	1849.1	154.6	32.1
Asbestos	77.9	41.5	8.6
Other Crude Minerals	510.0	26.9	5.6
Emerald	-	14.3	3.0
Potassium Salts (Natural)	55.0	8.0	1.7
Magnesia (Dead Burnt)	7.3	3.8	0.8
Borax	9.1	3.1	0.7
Fluorspar	8.8	1.3	0.3
Salt Rock	37.7	1.2	0.2
Magnesia (Fused)	2.1	1.1	0.2
Others	-	4.9	1.0

Year: 1985-86, Value Rs 481.0 Crores. Source: M.S. Division: Indian Burea of Mines

(c) methods to utilise lower grade phosphates in the existing fertilizer manufacture and agricultural techniques. The importance of these efforts become highlighted when we note that the projected requirements of phosphatic fertilizers as P_2O_5 are 3.2 and 4.5 million tonnes for the years 1990 and 1995 respectively. A further increase will be necessary for achieving th envisaged food production of 240 million tonnes by the year 2000 AD.

6   Some Concepts for Minerals Policy for India

We shall now outline the various features which should be incorporated in the national minerals policy. In this context it was felt that it would be useful to study and evaluate the mineral policies adopted by various other countries. (See *Strategic Minerals; Vol.1. Major Mineral Explorating Regions of the World* Vol.2. Major Mineral Consuming Regions of the World; by W.C.J.Van Rensburg, Prentice-Hall 1986). It emerges that, among the various countries of the world, Japan is outstanding in this respect and has a definitive and comprehensive minerals policy. We shall briefly describe Japan's minerals policy features before focusing on those of our own.

6.1   Features of the Minerals Policy of Japan

The national minerals policy of Japan comprises eight elements. These elements are listed and discussed below:

Development of domestic resources and maintenance of domestic production

To aid the maintenance of domestic production, even when it is uneconomical, an emergency finance fund has been created. When the prices of metals (being mined and extracted) fall below a certain minimum level, mutually agreed upon by the government and private mining companies, loans at low interest rates are provided to the companies from the emergency fund. On the other hand, if the price of metal exceeds the present maximum value, the companies contribute the extra money to the emergency fund.

To aid exploration within Japan, a 3-level national exploration programme has been implemented. At the level 1, geological assessment is made over large regions and the cost for this exploration is fully borne by the Government. Exploration surveys in the areas, found to be most promising by level 1 exploration, constitutes the level 2. For level 2 exploration, 80 per cent of the cost is borne by the government and 20 per cent by the interested private company. Detailed exploration on the very specific areas constitutes level 3 and the mining company concerned bears all the expenses for this level of exploration.

Development of foreign resources

Japan gives low interest loans, and credits to foreign countries for developing their resources. Invariably, the necessary technology for mining and extraction of the ore is also provided by Japan. In return, Japan receives a part of the final product.

Japan also signs long term purchase contracts (LTC) with foreign firms or countries for purchase of ores or metals. The average period of these contracts is around 12 years even though some of the LTCs extend upto 25 years. All these contracts guarantee a certain quantity of the mineral/metal every year but the prices are not usually fixed.

Stockpiling

60 days supply of metals (e.g.. nickel, copper, tungsten, aluminium, zinc and cobalt) is stockpiled for emergency use. The private industry is responsible for this stockpiling though they receive government guaranteed loans.

Deep seabed mining

The Japanese government has actively encouraged deep seabed mining from 1970 onwards. It is anticipated that Japan's advanced technological capability can be exploited for gaining a lead for that country in the promising area of seabed mining.

Changing national economic base

Since Japan is dependent totally on imports for all its raw materials, it has taken concrete steps to transform the economic base from the *energy and resource intensive economy to the knowledge and service intensive economy*. Towards this purpose, the government has identified the following areas as thrust areas along with the materials that will be required in these areas (Table 24):

Table 24

Japanese Thrust Areas and the Required Materials

Thrust area	Required materials
Aerospace	Ti, Co, Be, Ni, V
New energy sources	Si, Zr, Be, U
Information processing	Si, Ga, Ge, Ta, Cu, Pt-Gp
New basic materials	Clays, petrochemicals, Co, Mn, Cr, V, Ti, REE, Si
	Mn, Cr, Mo, V, Ti, W, Ni
Speciality steels	

Government - Mineral Industry Interaction

This area includes

● tax policy
● tariff on imports
● Government aid for depressed metals and minerals.

Research and Development Funding

Japan has mounted a major R&D effort especially in the thrust areas.
The distribution of Government funds for R&D is as follows (Table
25):

Table 25

Distribution of Japanese Government Funds for R&D

Establishment	Per cent Funding
Private Industry	70
Universities	18
Government Institutions	12

Substitution and recycling

Since Japan imports most of its resources, recycling, (and to some extent substitution) is given an important place in the minerals policy itself.

It is obvious that the fact that Japan imports most of its minerals has clearly biased its minerals policy.

6.2 Elements of Indian Minerals Policy

The first step in the evolution of a national minerals policy is the classification of all the metals as (a) abundant; (b) insufficient and (c) strategic on the basis of existing and proven reserves of the mineral and its rate of depletion. This requires the following steps :

● Identification of potential areas for a mineral based on geological criteria, i.e., geological base, gravity and magnetic maps.

● Intensive exploration of these areas using the latest techniques.

● Based on these explorations, constantly update and assess data of ore reserves, with greater emphasis on resources of immediate relevance.

● Estimation of the current rate of depletion of the deposit and its projected rate of depletion in the next 50 to 100 years.

● Based on the above data, classify the metal (ore) as abundant, medium or insufficient.

● In addition, examine whether the mineral is of strategic importance to the country.

The above classification needs to be constantly updated based on the latest available data, since the classification of ore reserves into various categories is dynamic and everchanging as illustrated in *Fig.20.*

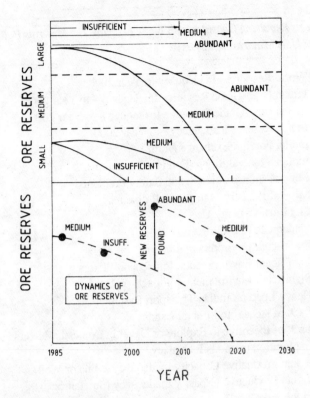

Fig.20 Classification of ores.

Once the above classification is available, the minerals policy ca be developed on the lines described earlier in this report, for each of the various categories noted above, in terms of the specific features demanding attention when the total picture from ore-to-product is considered. It is to be reitarated that the total picture with regard to the mineral under consideration should be taken into account while framing the mineral policy. For example, Table 26 gives a list of the aspects that needs to be considered and evaluated for each mineral as envisaged by U.S mineral industry.

In the following sections we describe the main elements of the policy for metals whose reserves are (a) abundant, (b) insufficient and for (c)

Table 26

Aspects Requiring Consideration While Framing The Minerals Polocy (vide USA Minerals Analysis)

1. World Resources Inadequacy to Year 2000.
2. Indian High-Grade Resource Inadequacy - to Year 2000.
3. Indian Low-Grade Resource Inadequacy- to Year 2000.
4. Indian Reserve Inadequacy - to Year 2000.
5. Indian Foreign Exchange Drain.
6. Indian Vulnerability to Foreign Disruptions.
7. Mineral Industry Health and Safety Problems.
8. Mineral Industry Man power Problems.
9. Significant Energy Use.
10. Inadequate Recycling.
11. Significant Environmental Impacts on Air.
12. Significant Environmental Impacts on Waters.
13. Significant Environmental Impacts on Land.
14. Heavy Load on Indian Transport System.
15. Lack of Access to Mineral Lands.
16. Lack of Incentive to Explore or Develop Domestic Resources.
17. Exploration methods Inaequacy.
18. Indian Productive Capacity Inadequacy (Mine or Well).
19. Indian Productive Capacity Inadequacy (Smelter or Processing Plant).
20. Inadequate Substitutes.
21. Inadequate Recovery from Current Extraction & Process Processing.
22. Inadequate Recovery of Byproducts and Coproducts.

strategic metals. The minerals policy for medium category metals will incorporate more or less the same features of the policy for abundant and insufficient categories and hence will not be discussed separately. The essential elements are illustrated in *Fig. 21*.

Abundant Reserves

The main elements of the minerals policy for metals with abundant reserves are listed:

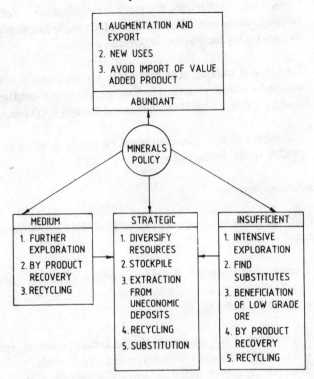

Fig.21 The essential elements of minerals policy.

● Augmentation of ore production and improvement of mining efficiency.

● Aggressive minerals export policy to derive maximum benefit of increased ore production. As far as possible, the export of value-added products, based on these abundant minerals, has to be attempted. In the case of virgin mineral export, the buyer nation should preferably be one which provides, in return, a commodity we are deficient in.

● Cartels can be formed with less developed countries which also have abundant resources of the same mineral, with a view to getting higher and stable prices for the ore.

- In areas where our mining and extraction technology is well developed, technical collaboration with LDCs which do not have the technology but the reserves is possible.

- The use of abundant metals, in either new applications or as substitutes for another metal in which we are deficient, should be encouraged by a system of incentives and through R&D funding.

- The import of such abundant metal especially as value added product should be discouraged.

Insufficient Reserves

For such metal, the minerals policy should incorporate the following essential features:

- Exploration based on a well-formulated strategy should be carried out. Target areas for exploration should be identified on the basis of geological criteria.

- Exploration of the seabed within the Indian territorial rights should be speedily and intensively carried out and areas of mineral potential (eg. copper, nickel and cobalt) should be earmarked for exploration.

- Beneficiation of low grade ores should be considered.

- Target and incentive-based recycling of the metal should be encouraged.

- The possibility of obtaining the metal in question as a by-product during processing for some other metal should be explored.

- Import the metal/mineral as part of a cartel or in return for exporting our abundant mineral.

- Discourage the consumption of the metal by finding suitable substitutes through R&D.

● Stockpiling of the mineral/metal to meet our requirements during emergency.

Strategic Minerals

In the case of strategic minerals/metals the following policy needs to be adopted :

● Exploration should be intensified to identify indigenous sources either as primary ores or as by-products.

● Diversify foreign sources of supply so that we are not unduly dependent on one country or countries from one block.

● Stockpile the metal in required quantity to meet the demand during emergency periods.

● Extraction of the metal from even lean, uneconomic deposits should be encouraged by the Government by offering subsidies or soft loans.

● Target and incentive-based recycling programmes should be initiated on a top priority basis.

● Long term R&D plans should be drawn up for ultimately substituting the strategic metal with some other material more readily available.

7   Summary of Recommendations

● It should be emphasised that the minerals policy should vary depending on whether the mineral falls in the abundant, medium, insufficient or strategic category. The above categorisation is dynamic and hence requires a constant updating of data on the reserves position, rate of production, rate of consumption, projected demand and international situation.

● It is suggested that from the abundant category, the minerals to be

731

exported can be identified on the basis that our known ore reserves should last roughly for the same period as the reserves in countries which have abundant reserves of the same mineral. Thus, if the acceptable levels of annual ore production arrived at on the above basis far exceeds the current and projected demand of the mineral, it can be targeted for export. Such an analysis, based on the currently available data, suggests that the production of bauxite (aluminium), chromite (chromium), ilmenite (titanium) and monozite (rare earths) could be augmented for purposes of larger export. Similar analysis should be carried out in the case of non-metallic minerals.

● Once a mineral has been targeted for export, the consideration that must dictate the choice of the countries to which the mineral should be exported, as well as the form in which it should be exported, requires imaginative analysis. Minerals represent a national asset and further augmentation of their production and export should result in maximum return benefits. Invariably the attempt should be to export value-added products. If export of ores themselves is made, the aim should be to obtain a locally needed raw material in return or to gain a strategic advantage for the country.

● The minerals policy should be developed into an integrated policy, covering the range of activities from exploration of the ore reserves, mining, beneficiation, establishment of extraction, refining and finally processing to relevant products (*Fig. 22*). Thus, a comprehensive analysis should be carried out for each mineral with regard to each of the above activities in the ore-to-product chain and the minerals policy, should then aim at removing the bottlenecks. Such a policy covering the range of activities from ore to product, requires linkages to be established among several organisations. In this endeavour although the DAE model is specific to strategic atomic minerals and nuclear materials, it is illustrative (*Fig.23*).

● Case studies on abundant minerals like iron ore (steel), ilmenite (titanium) and monozite (rare earths) point to contrasting pictures. Iron and steel bring home, against the backdrop of abundance of indigenous resources, the necessity for improving the quality of production and matching the production profile (with regard to shape

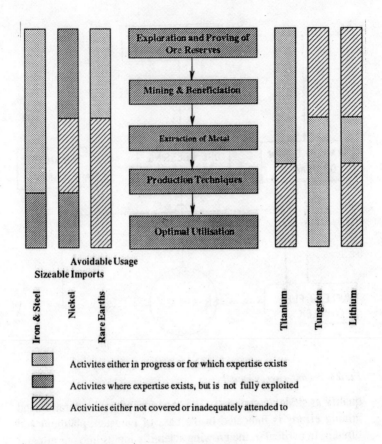

Fig.22 Integrated approach to minerals policy.

and type of steel) with the consumer demand, so that imports are minimised. Rare earths illustrate the requirement of compound separation and metal extraction technologies while, in the case of titanium, a policy directed towards use of titanium in the power generation and other (e.g. chemical) industries along with the setting up of a large scale metal production facility is indicated.

● As per the definition adopted for defining strategic materials, tungsten, the platinum group of metals, lithium, cobalt and nickel

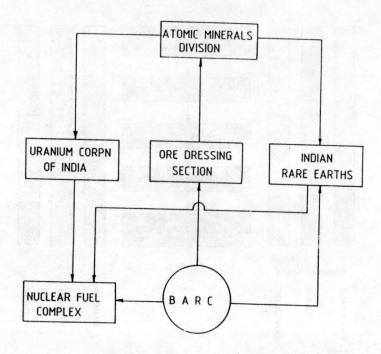

Fig.23 Atomic energy model.

qualify as strategic materials. The need for intense exploration and mining efforts is indicated in the case of tungsten, platinum and lithium. In particular, the growing scientific inputs into our mineral exploration activities need to be considerably encouraged. Cobalt exemplifies the gap that exists for recovery of this metal as a by-product from copper and zinc production. In contrast, in the case of nickel, the critical bottleneck is the development of an efficient and economic method to extract nickel from the laterite ores. Till such an option becomes viable, substitution and recycling need to be accorded such priority in the minerals policy that an impact is made through reduction in imports.

● The above recommendations, though illustrated through examples pertaining to metallic minerals, apply equally to non-metallic minerals. A case in point is rock-phosphate, in which case indigenous efforts for exploration (for high grade variety), beneficiation and even development of methods to use low grade material are called for to reduce the growing volume of import.

● Although there are a number of agencies and laboratories dealing with the various activities pertaining to exploration, beneficiation, mining, export/import of ores, as well as product development, there is a need for a single apex body which can take major decisions as suggested above, especially in respect of strategic minerals and those related to new materials of emerging significance.

15
Building Materials

Because of the stupendously high prices of building materials, nearly sixty per cent of the urban households cannot afford a cement house. The situation in rural areas is even worse; sixty per cent of the rural population will continue to have mud and thatch houses even in the year 2010.

Contents

15

Building Materials

Preamble

The congenial surroundings of a home are necessary for an individual's growth and also for national economic growth. The provision of urban and rural housing is a complex issue with the question of availability of housing finance, land servicing, high price of construction and housing delivery system being the critical constraints, specially in the urban areas. A recent evaluation, of the Sites and Services Project for the urban poor by the World Bank, concluded that while the concept was a valid one, replicability remained elusive largely owing to erroneous judgements on appropriate affordability estimates. The stupendously high prices of building materials have resulted in a situation where nearly sixty per cent of the urban households cannot afford a cement house. The situation in rural areas is even worse; sixty per cent of the rural population will continue to depend on mud and thatch for housing for at least another two to three decades. Whereas spectacular advances have taken place in the general area of application of new materials in sectors such as transportation, communication, energy etc., there has not been much progress in building materials.

It is here that S&T can play a major role. An examination of the present production of conventional building materials in India shows that there is a major shortfall in practically all areas. A significant breakthrough in the development of low cost and alternative building materials is necessary if we wish to fulfil the resolve of providing shelter for all our people.

The subject of building materials has acquired an importance today largely due to the fact that a number of initiatives launched by the Government in the housing sector are likely to give a fillip to housing construction, which, in turn, will create shortage of building materials, and push the prices even further. In this report, the availability of diverse raw materials, energy and infrastructural facilities and also various socio-economic and techno-economic factors, have been examined. In the light of these, alternative materials and processes have been identified and suggestions have been made on the possible strategies to be adopted for research and development in the areas of building materials. ■

1 Overview of the Housing Sector and Conventional Building Materials

1.1 Housing

The housing sector has played a peripheral role in the planning process. The national investment in housing has declined drastically from 35% in the First Five Year Plan to 7.5% in the sixth. Indeed, during the Seventh Plan, the Government could make an investment of only about Rs 2900 crores. During the last one year, however, three major events have occurred, which suggest that the Government has initiated processes which will reallocate national resources to the housing sector. These are:

● The formal adoption of a National Housing Policy Document (The first one since Independence)

● The publication of the final report of the National Commission on Urbanisation

● The creation of a National Housing Bank

It therefore appears that the government has realised the importance of housing as a source both of employment generation and of capital formation and wealth.

With the creation of a National Housing Bank and the major entry by insurance companies and commercial banks in the housing finance sector, the effective demand for housing will increase, thus creating a further housing shortage, which already has a heavy backlog. It has been estimated that rural households rose by 140 per cent in 1961-71 and a further 14.9 per cent in 1971-81. Urban households increased by 147 per cent and 40 per cent in the same time span. The data for the current decade are not yet available. However, the net result of this growth has been a massive housing shortage, which has been estimated to be anywhere between 25 to 30 million units, out of which 70 per cent are in the rural sector. The non-availability of land and finance are considered to be the major bottlenecks, but it will be erroneous to believe that land distribution by the Government, and the availability of easy housing loan on soft terms by Government/financial institutions, will solve this problem. The shortfall in the supply of conventional building materials is itself a great hurdle in the achievement of planned targets. A massive programme of investment on housing will accentuate the shortage of building materials. Except for cement, which is manufactured in the organised sector, hardly any reliable data are available on the capacity and production of the building materials industry. Table 1 shows the estimated present and future (by 2000 AD) production and shortfall in different manufactured building materials, if the targets in housing are to be achieved. It may be noted that these manufactured building materials take care of only 15 per cent of the population, which stays in urban areas and can afford pucca houses of cement construction. A functional classification of different building materials is given in *Fig. 1*.

1.2   Conventional Building Materials

From the point of view of raw materials and manufacturing technology, conventional materials can be divided into two categories:

● Naturally occurring materials with or without modifications

● Processed and manufactured building materials.

743

Table 1

Production and Shortfall in Conventional Building Materials in India

Material	(Units)	Year 1988-89		Year 2000 AD	
		Production	Shortfall	Production	Shortfall
● Bricks	(million)	63,000	17,000	1,00,000	25,000
● Cement	(million tons)	46	0.00	100	0.0
● Coarse Aggregate	(million cu.m.)	100	15	150	30
● Lime & Cementitious Products	(million tons)	4.0	2.0	7.0	3.0
● Ply wood	(million sq.m.)	66	60	100	65
● Particle Board	(tons)	80,000	28000	90,000	33,000
● Fibre Board	(tons)	1,00,00	30,000	1,10,000	33,000
● Plastic Building Materials	(tons)	1,59,400	42,000	4,26,200	Not possible to estimate

Source: CBRI estimates.

Naturally Occurring Materials

These are traditional materials evolved over the centuries through successive development and experience. These materials are ideally suited for rural housing.

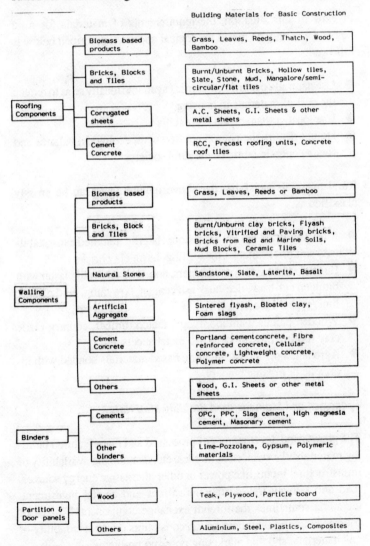

Building Materials for Basic Construction

Roofing Components	Biomass based products	Grass, Leaves, Reeds, Thatch, Wood, Bamboo
	Bricks, Blocks and Tiles	Burnt/Unburnt Bricks, Hollow tiles, Slate, Stone, Mud, Mangalore/semi-circular/flat tiles
	Corrugated sheets	A.C. Sheets, G.I. Sheets & other metal sheets
	Cement Concrete	RCC, Precast roofing units, Concrete roof tiles
Walling Components	Biomass based products	Grass, Leaves, Reeds or Bamboo
	Bricks, Block and Tiles	Burnt/Unburnt clay bricks, Flyash bricks, Vitrified and Paving bricks, Bricks from Red and Marine Soils, Mud Blocks, Ceramic Tiles
	Natural Stones	Sandstone, Slate, Laterite, Basalt
	Artificial Aggregate	Sintered flyash, Bloated clay, Foam slags
	Cement Concrete	Portland cement concrete, Fibre reinforced concrete, Cellular concrete, Lightweight concrete, Polymer concrete
	Others	Wood, G.I. Sheets or other metal sheets
Binders	Cements	OPC, PPC, Slag cement, High magnesia cement, Masonary cement
	Other binders	Lime-Pozzolana, Gypsum, Polymeric materials
Partition & Door panels	Wood	Teak, Plywood, Particle board
	Others	Aluminium, Steel, Plastics, Composites

Fig. 1 Classification of building materials.

745

More than half of the population would continue to live in villages even upto the end of the century and, because of the population growth rate of approximately 2.3%, the number of people living in rural India will be even higher than what it is today. Therefore the rural housing problem is likely to worsen.

In view of this situation, the requirements of materials for rural housing can only be met if the logical approach described below is followed.

- Use naturally occurring materials available locally so as to reduce the cost of materials in housing
- Use local craftsmen and their skills
- Do not disturb the natural environment, ecological balance and socioeconomic fabric of the rural society.

Rural materials satisfying the above requirements can be grossly classified as:

- Mud, sun-dried bricks, local stone, laterite, rammed earth stabilised soil bricks/blocks for walling, burnt clay bricks
- Lime *surkhi* mortars and plasters, non-erodable mud plaster with bitumen cut-back, rice-husk ash cement, cow dung, mud plaster, lime/gypsum plaster
- Wooden ballies, bamboo, twigs, thatch, timber, country made clay roofing tiles, stone slabs, metal sheets, etc.
- Agro and agro-industrial waste based materials bonded with inorganic or organic binders.

Processed and Manufactured Building Materials

The growth and development of processed and manufactured building materials like steel, cement, tiles, etc. demand the availability of infrastructural inputs like power or other alternative energy sources, road and railway transportation facilities and capital investment, including sometimes the foreign exchange component. This is, of course, assuming that there are no constraints on the availability of raw materials, and that marketing poses no problems.

Considering the infrastructural constraints on fuel/power, transportation and the perennial resource crunch, the preference should be given to the building materials industries which:

- avoid the use of energy or use comparatively lower energy for manufacturing materials of equivalent functional efficiency
- require lower power consumption per unit of production
- have lower capital investment per unit of production
- do not require elaborate and costly pollution control equipment
- do not have long gestation periods
- have consumption centres within an economic transportation distance.

Based on the above over-riding criteria, we have made suggestions in this report on innovative technologies and new building materials.

2 Cost Effective and Innovative Alternatives for Building Materials

2.1 Roofing Components

Roofing represents one of the major components of a building with reference to its cost and is equally important for the durability of the entire structure.

The roof can be generally subdivided into two components: the roof structure and the roof covering. The basic purpose of the roof structure is to withstand the loads transmitted from the roof covering. The roof structure generally consists of materials which have good flexural strength. It is, however, possible to have roofing systems which perform the dual functions of the roof structure and the roof covering. The performance requirements expected of a roof covering are that it should not leak, it should not erode under rain impact, it should provide good insulation from heat or cold winds, it should not be easily attacked by termites/insects, and it should have high strength to weight ratio.

2.1a   Roofing for Rural Houses

The choice of roofing material for rural houses is aggravated by the constraint that there are hardly any durable local materials which can satisfy all the performance requirements listed above.

Among the traditional roofing materials for rural houses, bamboo, thatch, clay tiles and slate/stone are among the most commonly used. More durable, and also more expensive, are the Mangalore tiles, asbestos cement, and corrugated sheets made of galvanised iron.

Bamboo plays an important role in house building in rural areas. Its use as a structural support to the roof covering is almost universal. In many places, it has been traditionally used for reinforcing mudwalls as well. The following are the advantages of using bamboo in the building sector:

● It has a considerable strength
● It can be harvested in five to six years in contrast to fifty to sixty years required for hardwood species of timber
● Its processing after harvesting is relatively easy and it does not require the elaborate machinery used in timber processing
● It can be grown almost anywhere close to a source of water.

While the importance of bamboo as a building material cannot be underestimated, it is also necessary to look at the alarming rate of depletion of bamboo resources in the country. The numerous paper mills that have come up in the last few decades have severely eroded the natural bamboo forests in the country. Today the paper industry is able to get bamboo at one fifth of the open market price! This use in the paper industry has pushed the traditional use of bamboo in house building into a secondary activity. Bamboo and other timber-like materials should be given for housing on a priority basis, and also at cheaper rates.

Although the excellent strength of bamboo leads to its extensive use in roof structures, its durability is often less. It is generally subjected to following kinds of deterioration:

- decay due to fungal attack in exposed situations
- attack by powder post beetles which often tend to colonise whole bamboos and decimate them
- attack by termites.

A simple and environmentally sound technique to reduce insect attack is to leach out the sugars by soaking the bamboo in water for about a month prior to use.

A satisfactory treatment for bamboo is desirable if its use in building construction is to be rendered free of problems. Several well known methods of timber treatment can be employed, but their use in rural areas appears to be doubtful due to the high cost involved and also the non-availability of various treatment chemicals. The sap displacement treatment, using copper sulphate, sodium dichromate and boric acid solution, looks promising.

While bamboo represents a valuable local material for the roof structure, the soil is the obvious local material for the roof covering. It has been used as a roof covering in its natural form, in traditional houses all over the world. However, the traditional soil roof covering is so thick and heavy that it pushes up the cost of the roof structure in a significant manner. The situation can be improved by using a thin layer of stabilised soil. Since stabilised soil is not waterproof, an independent waterproofing method is needed. The use of a layer of polyethylene sheet can be useful in this context. The stabilised soil covering will also generally need a protective coating to prevent erosion due to rain water impact and flow. In this system leakages can still take place, due to accidental holes and cuts in the polyethylene sheet, and, secondly, there is no composite structural action between the split bamboo and the stabilised soil layer. Water proof roofs can also be built by splitting lengths of bamboos into two semi-circular sections and then overlapping these sections like tongue and groove joints.

Thatch is a low cost roof covering material used widely in the country. A thatch may consist of reeds, sugarcane leaves, coconut leaves, palm fronds, grass or straw left over after harvesting wheat, rice, oats, barley and other cereals. Thatch, however, has many drawbacks: it catches fire, leaks, harbours rodents and reptiles, is attacked by termites and needs regular replacement. The published literature contains a method of fire-proofing thatch by impregnation with a solution of ammonium phosphate, which is a common fertilizer.

A study conducted by RRL, Trivandrum on coconut leaf thatch, which is widely used in Kerala, has suggested two probable solutions for extending its life. The first is to use a combination of a fungicide and a water repellant surface coating, having plasticising and anti-oxidant properties, and the second is the use of a fungicide that can be polymerised *in situ* to give a water repellant surface coating. Laboratory studies have indicated that the life of coconut leaf thatch can be extended from one year to four years by using copper sulphate as a fungicide and cashewnut shell liquid (CNSL) as a water repellant coating. Phosphorylated CNSL was, further, found to be an excellent non-leachable fire retardant for coconut leaf thatch. Replacing CNSL by phosphorylated CNSL, alongwith copper sulphate, gives a fire-retardant coconut leaf thatch with good longevity.

CBRI (all abbreviations used in the text are listed in Appendix-1) has also worked on the preservative treatment of thatch. The method developed involves the use of bitumen stabilised mud plaster and *gobri* on top of thatch with a final lime wash. This makes a roof heavy, leading to heavy supporting frames. But since it is a cheap process, it may be worthwhile to try this approach. Another option is to use the so-called tarfelt-cum-thatch roof. Tarfelt is a fairly cheap material. With a few rolls of tarfelt, supported by closely spaced thin wooden membranes, one can make a light roof. The layer of thatch becomes very thin, because the tarfelt layer provides the water proofing. The thatch on top can also be treated to further extend its life. ASTRA is conducting experiments on such membrane roofs.

The potter's tile is an age old material for producing roof coverings. Such traditional clay products need to be upgraded. Due to the low productivity and the thermal inefficiency of the production processes, these products are almost as expensive as urban products like Mangalore tiles. Potters need to be given training in new techniques like pressing, fast firing, etc. This will enable them to make new products such as Hourdi blocks, hollow clay pipes, burnt clay flooring tiles, water proofing tiles etc., which cannot be made by using conventional potter's wheel.

Mangalore tiles make excellent roofing material, but they are heavy and expensive. Still they have been universally accepted in the rural areas of South India and are displacing many other traditional forms of roofing. However, the production of these tiles is carried out in a large scale in *urbanised* areas and this *accentuates the flow of resources from villages to cities*. The production is also mechanised and relies on electric power for the energy needed for pugging, moulding and conveying the tiles.

The corrugated sheets of galvanised iron (GI) and asbestos-cement are also widely used on account of their durability. These products, besides being expensive, suffer from various shortcomings. GI sheets become uncomfortably hot during summer and cold during winter. Also, they look ugly when rusted and can be hazardous during a hurricane. The present practice of using asbestos-cement products must be stopped since asbestos fibre is carcinogenic.

Red mud plastic sheets based on PVC appear to be promising, although the high price combined with the recent evidence of danger-ous levels of smoke density in case of a fire, are obviously limitations at the moment.

Corrugated roofing sheets from coir wastes/wood wool- cement/asphalt/paper need to be propagated. Some of the timber like prod-ucts, listed in Table 3 (on page 764 of this report) can also be used for roofing components.

The rural poor will continue to use bamboo/timber roof frame with covering of thatch, reeds or country tiles. The construction of small and medium span nail jointed timber trusses may be propagated. The utilisation of bamboocrete and reed boards for roof and wall panels may also be explored.

An approach, which involves a prefabricated framework supporting a roof, in which the villagers are able to infill the frame with locally available material needs to be tried in rural areas. The design must be aimed at encouraging self-help by involving a minimum of skilled-labour, and allowing flexibility in terms of size and materials to suit local conditions and preferences.

Reinforced brick panels and stone slabs are other good alternatives which need attention.

2.1b   Roofing for Urban Houses

Reinforced concrete has been universally accepted as satisfactory roofing for urban areas under widely different climatic situations. However, the cost of its construction is such that it is inaccessible to the vast majority of the urban population. An analysis of typical reinforced concrete slab shows that the available load carrying capacity of the slab is only about 53%, the remaining 47% is the weight of the slab itself. It must also be noted that about half the concrete in the slab (which lies below the neutral axis) does precious little in resisting the stresses induced by various loads. This is on account of the fact that concrete is rather weak in tension. It is thus clear that the reinforced concrete roof slab, as is normally constructed, is quite inefficient. Its structural efficiency is only about 53% .

Secondly, the very process of constructing, or putting the elements of the roof together, is very often an inefficient process, thus increasing the cost even further.

In the urban context, the concept of prefabricated roofs, which use reinforced brickwork or panels, with extruded hollow tiles, offers a very cost effective alternative to reinforced concrete. This technology was in use in the early forties (as is borne out by the old Bombay PWD Handbook). Unfortunately, the indigenous development along these lines was neglected due to the attractions of reinforced concrete. Reinforced brickwork has already been tried experimentally both by CBRI and ASTRA.

Fibre reinforced cement tiles, concrete roof tiles and ferrocement roof can be other alternatives. Concrete roof tiles are likely to enter the Indian market very soon. In fact a small unit has just started production near Pune. The prices are very competitive when compared to those of Mangalore tiles.

Precast RCC channel/cored slab/funicular or waffle shells/battens and hollow block construction/hyperbolic shell/planks and joist systems/folded plate roof/thin ribbed slab/L-pan roofing units and light weight concrete panels/cellular units need to be propagated mainly for mass housing schemes. Some of them can also be used for flooring. These have advantages such as economy in steel and cement consumption, lower deadloads in design, better quality control and faster rate of construction.

Fibre reinforced polyester (FRP) sheets are slowly penetrating the urban market. There is a need to bring the prices down so that their use can be enhanced. FRP conversion products manufacturing does not require large fixed capital and is a labour intensive activity concentrated in the unorganised sector. There is a need to give fiscal concessions to such products.

2.2   Walling Components

2.2a   Walling Components for Rural Houses

The most common forms of wall construction techniques in the rural areas are :

● Rammed earth - walls are built after pressing or ramming a lot of earth together
● Cob and adobe - hand laid or moulded walls in mud, a system which gets eroded by rain and animal attack and, therefore, needs to be protected
● Wattle and daub - a biomass mesh woven out of bamboo, twigs or branches, which is plastered over with mud on both sides.

The last technique needs to be improved considerably; in fact, double storey structures in Peru are constructed with this technique. Assam and the north east too have high quality infill panel buildings constructed using this technique. The major problems here are the decomposition of the biomass and the erosion of the mud because of rain.

Information on African, Asian and Latin American practices contains references to the use of natural additives, such as animal dung, extract from banana stalk, cactus pulp or locus bean pods. When added to mud, these are claimed to improve the moisture resistance of the mortar used for joining the adobe bricks, or the stucco (plaster) used for protection and beautification of wall surfaces. Several natural fibres are often added to improve the cohesive strength of mud blocks.

CBRI has developed a non-erodable mud plaster by using a bitumen cut back emulsion which has been extensively tried. IIP & IPCL have recently developed a PVC coating to water proof mud-walls of the rural house. This coating has undergone limited field trials and its practical application in rural areas, and cost effectiveness, are yet to be established. Alternatively, a layer of jute reinforced polyethylene sheet, obtained by cutting empty fertilizer bags, can also be used as damp proof course.

The best way to use mud for rural house construction is to use compacted soil blocks in a mud mortar. Techniques of water proof coatings, based on animal glue and white wash have also been effectively used to protect mud block walls to prevent erosion by rain.

A simple upgradation of traditional wall construction techniques using better tools and some stabilisers can have a big impact in rural housing. Rammed earth with aggregate or puddled earth with aggregate are some of the techniques which only need simple moulds for construction.

2.2b   Walling Components for Urban Houses

Bricks, Blocks, Tiles

Today the burnt clay brick is the most accepted walling component in urban areas. It has been in use since 10,000 BC and the current practices and trends indicate that it would continue to be used for many more years in the future. Nearly 90% of the brick manufacturers in India use primitive kilns which are energy-inefficient (thermal efficiency < 20%). In addition, the process of brick manufacture generates high pollution, creating very serious ecological problems. The following steps are necessary to correct this situation:

● To design, develop and promote pollution free and thermally efficient brick kilns, which may also employ alternative non-conventional fuels.

● To adopt soft mud brick presses. A modified version of press developed by MERADO, Ludhiana,is now available in Maharashtra. These presses are known to work satisfactorily with all clays without much processing. Extrusion and wire-cutting machines usually work well only with plastic clay and require substantial investment in clay processing machinery and worksheds. However, more operational experience and feedback would have led to a better promotion of soft mud brick presses.

● To promote the utilisation of fly ash, coal-mining waste, tailings etc., in the production of burnt clay bricks in order to minimise the consumption of good quality clay (see Table 2).

● To improve the quality of bricks, blocks and tiles made from inferior soils, such as black and red soils, by the introduction of scientific practices in processing and firing.

● To investigate the use of agricultural waste, such as rice husk, coir dust etc. as potential additives to the brick making clays, with the objective of reducing fuel requirements in firing, and of improving the quality of finished brick.

As a long-term policy, the use of burnt clay bricks should be discouraged to save energy. The average energy content of a burnt brick is about 1500 kcal/brick, and this energy is obtained from firewood or coal. A calculation of firewood consumption in south India shows that each brick consumes 0.32 kg. of wood. Thus a small

Table 2

Bricks and Blocks from Alternative Materials

- Coal-ash/carbonisation/mine tailings/beneficiation waste/washery waste
- Mine tailings/sludges/beneficiation waste (iron, copper, zinc, gold etc.)
- Red mud
- Waste from fertilizer industry (phosphogypsum)
- Mineral waste (flourspar, mica)
- Water works silts/wastewater sludge/activated sludge/river dredgings
- Calcium silicate/lime sludge
- Lignite waste/overburden
- Cement/asbestos waste
- Cinder (hallow cinder block masonry)
- Municipal refuse
- Fly ash / fly ash - gypsum / lime-sand-fly ash (bricks or cellular concrete blocks)
- Metallurgical slags
- Laterite soil (lato blocks)
- Rice husk ash
- Clay clinker/clay-sand
- Pre-cast crushed bricks

50 square metre house, in terms of bricks used, will need five tonnes of firewood. This makes brick housing difficult for a vast majority of people besides affecting the ecological balance.

The analysis of energy brick prices during the last 15 years shows that the energy and the cost of building materials are very closely related. The perennial scarcity of fuels has led to a steep escalation in the price of bricks.

The stabilised mud block offers an interesting energy efficient alternative to burnt bricks. The technique involves pressing of locally available mud into dense blocks in a manually operated machine. The performance of the mud block can be upgraded by the addition of stabilisers such as cement or lime (5% by weight). Although this increases the cost, the energy consumption is a small fraction of that in a burnt brick.

The use of cement stabilised soil blocks has been tried in the urban environment. Experiments have shown that unplastered external soil-cement block surfaces can perform satisfactorily, when the annual rainfall is in the range of 900 mm to 1000 mm. The use of lime, or a combination of cement and lime, is also quite feasible.

The use of stabilised soil for house construction has already been tried way back in 1948 in Punjab. 4000 rammed soil-cement houses were constructed for refugees around Karnal in Haryana. The project was backed by three years of earlier research at Lahore and at the Building and Road Research Laboratory in Karnal.

Recently, the ASTRA centre of Indian Institute of Science, Bangalore has developed two machines, *Astram and Itage Voth*, for compacting soil stabilised blocks. About 200 machines have been sold. A number of houses and factory buildings have been constructed using stabilised soil blocks made from soil dug out from the site. This concept not only leads to energy saving, but also helps in avoiding transportation of clay and finished bricks. This is an example of a promising technology, which needs to be propagated all over India.

Some other voluntary agencies like Development Alternatives have developed a soil block making machine, *Balram*, under the CAPART project. About 100 such machines have been marketed.

HUDCO is working on semi-hollow stabilised soil blocks to reduce the dead weight. They are also planning to provide a cement facing to soil blocks, at the time of manufacture, by keeping a separate compartment in a mould. This will avoid the need for plastering. Such ideas and techniques need urgent attention.

Prefabricated brick panels have been tried out for roofing. Their use in walling also increases the efficiency of the construction technology.

There has been some criticism about the use of clay bricks on account of their effect on the loss of soil. In the present socioeconomic situation the exclusion of clay products from building construction activity would appear to be rather impractical.

Sand is one of the neglected resources which can be used for the manufacture of walling components such as sand-lime bricks. However, the techno-economic viability of sand-lime bricks needs to be worked out in the Indian context. The advantage of using sand-lime bricks is that bricks can be obtained in different colours thus avoiding plastering, etc. Sand can also be used for the manufacture of tiles after pressing it with resins.

Stone

Many regions in India do not produce quality bricks; indeed, good bricks are unthinkable in our desert, mountainous and hilly regions. Consequently, wall construction in these regions is done either in the form of massive random-rubble masonry or by using expensive dressed-stones. Also walls are often erected in costly brick masonry by using bricks transported from distant places.

Stone block masonry which has been developed by CBRI, is an ideal walling system for all the stone abundant regions where bricks are either of poor quality or costly.

The technique uses 30% of irregular stone spalls and the remaining 70% of lean cement concrete to cast stone masonry blocks in simple steel moulds. The technique is not only economical but labour intensive too. The system is so simple that local unskilled labour can produce these at the work site. The spalls used may be either broken rock pieces or from river boulders.

Stone is a widely available raw material (for example : Cuddapah in Andhra Pradesh, Shahabad in Karnataka, Sandstone in Rajasthan). However, it has remained a neglected resource from the development point of view. It is one of the most durable material of roofing and walling. It is also widely used not only in rural areas but also in urban settlements in north and central India. However, its utilisation is decreasing because of the following reasons :

- Commercialisation through contractors of small quarries and the displacement of local artisans

- No improved extraction technologies and low remuneration to stone workers.

The extraction, processing and design of buildings with stone has to be investigated very carefully. This is a material, which can produce high quality, cheap and durable buildings as seen in many of our monuments such as forts, etc. However, an upgradation of the existing technologies is essential.

Prefabricated cement-concrete products/cellular concrete/lightweight expanded clay aggregates

Considerable progress has been made in advanced countries towards industrialised housing systems through prefabricated cement-concrete products. Today there is no distinction in the definition of building materials such as brick and a full ceiling-high prefabricated panel which has made erection of a complete house a very speedy process. In addition, tremendous progress has been made in the manufacture and use of light-weight cellular (or foamed) concrete and clay/slate based light-weight aggregate concrete.

Interestingly enough, in India, foamed concrete was first introduced in 1948 in Hindustan Housing Factory, New Delhi followed up by two other factories in Poona and Ennore. These units have been struggling to establish a good base in construction industry.

One of the main reasons for these products not finding acceptance in the private constructions is their high cost, which includes excise duties. Another one is the obsolete tendering system of Government departments. These materials do have certain advantages like light weight, less consumption of cement, less construction time, uniform quality, ease of operations, etc. There is need to encourage the use of such materials in the urban context, where most of the construction is done by third parties i.e. builders. However, the housing delivery system in India is characterised by relatively small 'builders' rather than large scale construction companies constraining the benefits that might be derived from economics of scale, modular construction and industrialised techniques. This is, in part, a consequence of the non-availability of large tracts of urban land for residential housing projects.

In the case of rural areas, it is not possible to introduce the industrialised building materials/prefabricated technology due to various adverse socioeconomic factors such as high cost, unemployment, etc., since most of the construction activity in the rural areas is undertaken by artisans and also on 'self-help' basis with the available local resources.

Even in the urban context mechanised operation throws people out of employment in a sector which is probably most labour intensive. Indeed in labour intensity it is next only to agriculture. What is appropriate for us is the intermediate level of prefabrication to be done at site which saves capital investment, energy, and transportation cost.

There is a need to develop prefabricated hollow units, which butt together and interlock at joints.

Attention also needs to be given to the development and production of artificial aggregates, such as sintered fly ash and bloated clay light weight aggregates. Some of the waste materials which can be used for the light weight expanded aggregate include fly ash, mine tailings, red mud, slags, coal wastes. etc.

2.3 Binders

Portland cement is today the choice binding material for use in construction. Currently, an efficient production of Portland cement is done in large centralised plants. The cost of production in such plants is reasonable. However, the cost of distributing centrally produced cement is higher and this more than offsets the lower production cost. In the final analysis, the local production of binders assumes significance due to the fact that at present each bag of Portland cement travels, on an average, a distance of 600 kms before reaching the consumer. This kind of a product can never be sold at a sufficiently low price in rural areas. Importance therefore needs to be given to small scale manufacture of alternative binders such as lime-pozzolana using local lime, burnt clay (*surkhi*) and rice husk ash or similar calcareous materials such as lime kiln rejects, mining wastes from bauxite, laterite and china clay mines.

Today nearly fifty per cent of high-strength Portland cement is used for secondary applications such as masonry joints, wall plastering, etc. This is clearly undesirable, since the strength potential of Portland cement is never fully utilised in such applications. There is an urgent need to manufacture different grades of cement. Production of blended cements, with optimal utilisation of pozzolanic materials and granulated blast furnace slag, should be encouraged. Formulation of guidelines for the use of blended cement requires urgent attention. One way to increase the use of pozzolana cement in secondary construction is to have a substantial differential in the price of ordinary Portland cement (OPC) and Portland pozzolana cement (PPC). Cement can be partially replaced by fly ash or reactive *surkhi* in mortar and concrete.

All the wet-process cement plants must be converted into dry process plants to considerably reduce energy consumption.

India has reserves of 69,000 million tonnes of limestone, which will last long. There is a need for setting up mini-cement plants, particularly in those parts of the country where limestone deposits are too small to sustain the demands of large size rotary kiln cement plants. The technology for mini-cement plants has been developed by RRL, Jorhat and NCB, and it is being successfully used at many locations. Some of the states do not produce enough limestone to promote lime and cement industries. In such a situation, various industrial wastes can be used for the production of low-cost cementitious binders and masonry cement. The use of the tailings from the beneficiation of zinc, copper and iron ores, fly ash, red mud, etc. for making ready-to-use masonry mortars, packed and supplied like cement, needs consideration.

The large-scale manufacture and use of hydrated lime from magnesian and dolomitic limestones, and the development and production of gypsum based binders, especially from the by-product gypsum obtained from the fertilizer, phosphoric acid and hydrofluoric acid industries should receive immediate attention. There is also need to develop thermally efficient and pollution free lime kilns of capacities ranging from one to ten tonnes per day.

2.4   Door Panels, Partitions, Claddings and Timber Like Materials

Timber forms 10 to 15% of the building cost. Timber is in short supply and its prices are rising very fast. Wood continues to be a major source of energy for cooking food and other non-commercial purposes. Very little timber is therefore left for use in building materials.

A complete ban on the use of timber and other woody materials in building construction is socioeconomically neither feasible nor desirable. As a short term measure, the present policy of import of timber will have to be probably continued. However, the duty structure should be such that the consumption is discouraged.

The biomass process is the cheapest, and the least energy-intensive. As a long term measure, plantation programmes for timber, bamboo and other species need to be developed on a priority basis with a twenty year time perspective. A thought should also be given to new propagation technologies such as tissue culture. Plantation should be done on waste lands by encouraging agro-forestry schemes. Assistance can be sought from agencies like NABARD.

In the case of rural housing, timber can be grown in blocks (dense planting) for local consumption. Timber, firewood and fodder plantation must be combined. Rural communities should be assigned community controlled lands exclusively earmarked for biomass production for local use. The social forestry programme should concentrate on this aspect. This can be achieved by leasing out the land to farmers' co-operatives.

All industrial wood and timber for high cost construction should be produced on private farmland. To render this a profitable business for farmers, a total ban on all subsidised supply of wood from reserved forests to these sectors, should be imposed. Even plantation on waste lands by private companies in the assisted sector may be thought of on a profit sharing basis. This is similar to what is being considered in some of the states for the paper industry. The technology mission on wastelands must incorporate production and social forestry components.

There is a need to improve the processing of timber. It is learnt that about 40% of timber is wasted in handling which includes operations in saw mills. Attention should also be given to local, fast-growing termite resistant indigenous/exotic species of trees, which need not be saw-milled. In the form of round polls, they will be useful for rural housing.

Similarly the productivity of timber must be increased. Consumption of timber/bamboo must be brought down by innovations in design such as frameless doors and windows, replacing of windows by jalliwork, etc.

More than 400 species of timber are grown in different types of forests in our country, only about half a dozen of them (primary species) are used for constructional purpose due to their natural durability. Secondary timbers are quite suitable after they are seasoned and chemically treated. About 40 species are easily available across the country. The emphasis should be to increase their consumption.

Concentrated efforts also have to be made in developing timber like materials and timber substitutes. These could be agro-waste based, plastics, reinforced cement-concrete or fibre reinforced composites

Table 3

Timber like Products (i.e. boards/panels) from Alternative Materials using Cement/Polymer Matrix

- Sisal
- Coconut/coir fibre
- Jute stalk/fabric
- Cotton stalk
- Baggasse
- Groundnut/Rice husk
- Saw mill waste/woodchips
- Bamboo mats
- Polypropylene fibres
- Coconut/palm tree leaves
- Rice straw/pulp blends
- Pine needles
- Inorganic fibre boards (rock wool)
- Cement/polymer composites
- Ferrocement door shutters
- Gypsum plaster boards
- Foamed boards - gypsum/vinyl chloride or expanded polystyrene or polyurethane products

(see Table 3). There is little development in advanced countries focused towards the use of agro-based materials in composites for low cost housing. The main reason for this trend is the availability of synthetic materials such as glass-fibre reinforced plastics and polyurethane foams at affordable costs in such countries. In India these synthetic materials are quite expensive and their prices are approximately three times the international prices. Therefore the approach for the development of low cost housing in India has to be different from the proven concepts demonstrated abroad.

There have been a few attempts in the country to develop composites from cellulosic reinforcements. However, the matrix material used in these composites has been an unsaturated polyester resin, which being hydrophopic, is not chemically compatible with cellulosic reinforcements. As a result, the interfacial adhesion is poor, thereby leading to the migration of moisture along the interface. This results in rotting and deterioration of the cellulosic reinforcements. The polyester resin is available at Rs 40/kg and therefore composites based on polyester resin are not cost effective.

NCL has developed composite panels from phenolic resins which are hydrophilic, and, therefore chemically compatible with cellulosic reinforcements. Secondly, since the phenolic resins are synthesised from a low cost indigenously available agro-based raw material, CNSL, the estimated cost of the resin would be in the range of Rs 15-20/kg. Thus it is expected that the composites may be cost effective. The panels have been made from woven bamboo mat reinforced composites, particulate filled composites and other cellulosic wastes. Adaptation of this in practice must wait until extensive field testing is done.

There is an urgent need for future R & D in this area, since a variety of matrices and agro-based cellulosic reinforcements can be used. These panels are useful for walling as well as roofing components. There can be separate products for external walls, where the aspects of weatherability and security are of prime concern, and those for internal walls, wherein light weight and sound insulation may be treated as the prime requirements. This would allow the use of a combination of inorganic and organic building materials in housing.

The use of synthetic materials such as fibre reinforced PVC tiles, in place of ceramic tiles, may reduce loads on the slabs. A combination of light weight internal wall panels and synthetic flooring could lead to smaller sizes of supporting beams and columns thereby resulting in conservation of RCC materials.

The use of galvanised iron tube structure, aluminium or steel channels in conjunction with composite wall panels may be considered for urban slum rehabilitation programmes.

The Government can play a major role here by altering the duty and raw material pricing structure of the thermosetting resins- which are inhibiting the growth of these alternative products.

Ferrocement is already used fairly extensively for water tanks. There is a clear possibility of using this material to replace timber in many applications like rafters for roofing, door and window frames, panels for shelf construction etc. The exsisting knowledge can be readily deployed towards this end.

2.5   Services

Some of the new materials/techniques mentioned below can bring down the cost of construction:

- Plastic pipes for water supply and drainage
- Plastic sanitaryware, fittings for doors and windows
- Fibre reinforced concrete manhole covers
- Ferrocement and plastic water tanks
- Bituminous materials and polymeric coatings for water proofing
- Single stack system of plumbing (one pipe system for carrying soil from toilets and waste water from kitchen, bath, etc.)
- Pour flush latrines of the *sulabh* type. A system of prefabricated sanitary core and solid waste disposal from fired clay needs to be developed and disseminated.

In addition to low cost innovative building materials, economy in architectural planning and structural design such as lower ceiling heights, cross wall construction, use of less concrete in foundations etc. can bring down the cost of construction considerably. Considerable work has been done on this aspect by CBRI, SERC and architects like Lauri Baker and others. This aspect has not been dealt with in this report. Work and ideas of such organisations/architects may be documented in the form of manuals, so that the knowledge can be disseminated for the benefit of community.

3   Use of Industrial and Agricultural Wastes as Raw Materials for Building Materials Industry

The foregoing discussion and a general survey of raw material availability indicates that the future research and development in building materials in India will revolve around the utilisation of industrial and agricultural wastes in a major way. Table 4 shows the present availability of some of the industrial and agricultural wastes in India and also their most potential use as building materials. Figs. 2 & 3 maps of India, which indicates the locations of some available industrial wastes, with potential locations for new building materials industries.

New materials industries based on the utilisation of industrial wastes have been successfully established abroad for a number of years. However, commercial exploitation of know-how of conversion of

Fig. 2 Availability of major industrial wastes for new building materials.

Table 4
Annual production of Industrial and Agricultural Wastes (1988-89)

Waste	Industry	Production (Million Tonnes)	Potential uses in Building Materials
Fly ash	Thermal power	35.0	Cement, brick, light weight aggregate, cellular concrete
Blast furnace slag	Iron and Steel Industry	10.0	Cement
Byproduct gypsum	Fertilizer, phosphoric acid, hydrofluoric acid	5.0	Cement, plaster of Paris, plaster plastic boards industries
Lime sludge	Paper, sugar, fertilizer, acetylene & tannery industries	4.5	Lime, cement masonry mortar
Cinder	Thermal power stations and railways using lump coal	2.5	Brick, aggregate
Mining tailings	Zinc, copper, gold, iron ores beneficiation	6.0	Calcium-silicate, concrete, brick, masonry cement plants
Red mud	Aluminium industry	2.0	Brick, ceramics, cement
Water work silts	Water works	10.0	Brick, cement, light weight aggregate

Table 4 (*Contd.*)

Waste	Industry	Production (Million Tonnes)	Potential uses in Building Materials
Lime kiln rejects	Lime kilns	0.4	Masonry mortar
Coal washery rejects	Coal washeries	3.0	Brick, light weight aggregate
Rice husk	Rice mills	25.0	Particle board, pozzolana
Coconut husk byproducts	Coconut based industries	3.0	Particle boards, roofing sheets
Jute sticks and bark	Jute mills	5.0	Insulation boards
Bagasse	Sugar mills	5.0	Insulation boards
Groundnut hulls	Groundnut oil mills	2.5	Particle boards
Rice straw,	Agricultural farms	90.0	Fibre boards
Wheat straw		33.0	
Corn cobs and stalks	Agricultural farms	14.0	Various building boards

Source: CBRI Estimates

wastes into building materials has not yet taken off in India. There are, of course, a few exceptions such as in the cement manufacturing industry, which is using slags, phosphogypsum etc. successfully with attendant advantages both in energy and cost reduction. One of the major reasons is that, in our country, the implementation of pollution control laws is very weak. Polluters go scot free. If all industries generating wastes are forced to dispose them off effectively, then their alternative uses would be automatically encouraged. Some of the other reasons for this state of affairs are:

Fig. 3 Possible locations for the manufacture of alternative building materials.

● high costs for transportation of industrial wastes from their sources to the building materials manufacturing plant site

● availability of industrial wastes far away from material consumption centres

● problems in scaling up processes, developed either on laboratory or pilot plant scale

● lack of engineering know-how in terms of equipment design, detailed engineering drawings and indigenous fabrication technology. This is in spite of the fact that, the engineering capability in the country is high and most of the problems could be overcome through proper coordination. One fact which has often been lost sight of is that the building materials industry is a process industry and mere availability of machinery is not sufficient

● market acceptability of the new materials.

Manufacture of building materials from industrial wastes can considerably reduce shortages of conventional materials, and, at the same time, produce new materials for different applications. In order to promote the use of industrial wastes, the following factors may be considered, and suitable concession, may be granted.

● waste disposal for building materials should be tied up at the project planning stage. User industry may be treated as an ancillary industry

● industrial wastes should be made available at no cost or at a nominal cost; otherwise they cause pollution and their disposal involves expenditure

● producers of byproducts and waste should extend all infrastructural facilities to the potential entrepreneurs. These include, for example, land, power, facilities for transportation, etc. in case the building materials manufacturing plants are established in adjacent or closeby areas

● building materials industries should be established in the joint or assisted sector to promote exploitation of industrial wastes as raw material resources

● subsidies should be granted on lines similar to those that are available for establishing industries in no-industry districts and in backward areas

● funds at concessional rates, and concession in other fiscal levies should be provided

● The subject of building materials should not be treated in isolation. It must be integrated with environmental/urbanisation/industrialisation issues.

As a first step, it is suggested that the Ministry of Urban Development undertake studies, with the assistance of NCB, CBRI, HUDCO and other agencies, on the following aspects :

● identification and availability of various wastes
● handling and transportation of various wastes
● trends in the utilisation of various wastes and the development of technologies for the manufacture of new building materials utilising wastes
● raw materials and energy conservation aspects of utilisation of wastes
● techno-economically feasible alternatives
● evolving appropriate specifications and codes of practice for application
● managerial and social aspects of disposal of wastes
● identifying further R & D and technological support needed.

4 Research and Development in Building Materials: Analysis of Past Failures and New Strategies

There is a major difference between the pattern of research in building materials in India and that in the western countries. Whereas in India only few government funded research institutions have been responsible for work, right from the concept to technology transfer, the scenario in the rest of the world is different in that the new innovations in the area of building materials/technologies have been achieved by industry. Today, in India also, R & D in related sectors like cement, paints, polymers and plastics is now being taken up by industry; yet there is no appreciation of the research needs in basic building blocks like bricks, stones, tiles, lime and timber products in the industry.

We have sometimes tried to copy the western world with disastrous results. A few automated brick making technologies imported in India did not meet the expected success and therefore it is necessary to have a fresh look into the development of clay-product industries suited to Indian conditions. India being a tropical country, clay is the only viable raw material because it provides a natural coolness (thermal insulation). The manufacture of terracota products i.e. bricks, tiles and roofing elements need attention from the point of modernisation.

The major problem of R & D in the building industry has been that there are very few centralised institutions, and field extension services are lacking in most cases. This can be illustrated by the fact that stabilised mud blocks were developed and used way back in 1948 in Karnal and this concept was further developed by ASTRA in 1974-75, but only during the last two years have they started penetrating the building industry and that also locally in Bangalore area only. Construction agencies/contractors all over the country are not yet exposed to this energy-saving technology; similarly stone block masonry, which has been used in experimental housing schemes of NBO, has yet to find wide acceptance. Another case of a potentially excellent technology not moving from the laboratory to the land is that of funicular/waffle shells developed by SERC, Madras. Although this technology for roofing/flooring was developed by SERC in the early 1960s, it has only now got some exposure due to the efforts of building centres at Quilon and New Delhi. It has also been found that very often no techno-economic evaluation of new products/processes has been done, no field data are generated, and there is poor awareness about new or upgraded technologies. The participation of R & D groups in field dissemination is negligible. This failure may be attributed to (a) lack of adequate R & D support to sort out problems of performance and (b) poor networking with technology delivery systems which could lead to a 'marketing' of the new concept. An important link in the chain of innovation consisting of 'architects - structural engineers - material manufacturers/suppliers - contractors - construction workers,' is often found missing.

Another basic problem regarding R & D in building materials in India is the fact that small scale, village level operations have hardly received any S & T inputs. The current research in national laboratories and universities has a tendency to concentrate on large, urbanised production systems. There is a need to transfer new and improved technologies to the small enterepreneur in rural areas. The selection of technologies has to be carried out in a participatory manner by involving the rural consumer and the local artisan/skilled worker. This is a time consuming but inevitable step that must be taken if simple and appropriate building technologies have to take root in the country. R & D in building materials has to be decentralised and it should be based on local resources. Economic evaluation should be done in different geoclimatic regions. There is also a need for housing extension agencies, on the lines of the agriculture extension agencies.

Although NBO was promoted way back in 1954 to promote demonstration, training and extension work, it has not been able to achieve the objectives. One of the major reasons is that it is not equipped to handle this gigantic task. It is learnt that the present technical staff of NBO at Delhi is only about 10 persons! With this kind of infrastructure, they need to be complimented for still being able to successfully disseminate information by way of publications and experimental housing schemes.

There is a need to integrate R & D with demonstration units for production as well as for application and training. Institutions must not merely develop new technologies, but must also make conscious efforts to monitor them in pilottest situations, to smooth out the angularities which will be invariably present in any nascent technology. A possible model is shown in *Fig. 4*. An institutional network for development and production of building materials, technologies and construction systems will be required. It is proposed that a Building Materials Development Board be set up by the Ministry of Urban Development as a non profit society. The major function of the Board shall be to:

● act as a repository of information on all types of building materials/technologies

● conduct technical evaluation, with the help of industrial experts, of various processes and technologies already developed and available for the production of different building materials and components

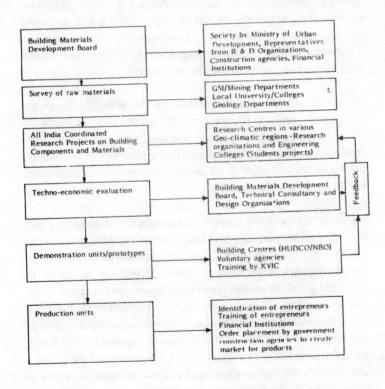

Fig. 4 *Model for institutional network for development and production of building materials/technologies.*

- act as a catalyst for technology transfer from laboratory to land

- assist entrepreneurs in covering the gap between research and commercial production

- promote and coordinate the appropriate institutions in different parts of country, such as building centres, which, in turn, would accelerate the production of innovative building materials and their application.

The financial requirements of the Board can be met by the National Housing Bank and other specialised housing finance institutions. The Board can also raise funds through membership of building materials manufacturers, entrepreneurs and others, to whom it will, in return, provide various types of services. The representatives of various R & D institutions, major construction agencies and financial institutions should have a say in the management of Board.

The Board will fund the survey study of raw materials in various regions of the country. These surveys can be done in association with GSI, state mining departments and geology departments belonging to the local universities/colleges. Based on the inventory of raw materials, the Board will identify the building materials and components which can be produced locally.

The Board is expected to support the All India Coordinated Research Projects (AICRP) on Building Components and Materials. Under AICRP, there will be various research centres in different geo-climatic regions of the country. These centres can be set up in engineering colleges where post-graduate students belonging to the civil engineering discipline will undertake research work. Based on the results pooled by different centres, on particular building materials/components, AICRP will make recommendations to the Board about the adoption of new technologies. The Board will then undertake techno-economic evaluation of new materials/technologies. If found feasible, the technology will be passed on to the building centres being promoted by HUDCO and other voluntary agencies for demonstration purpose/prototype development/training. Even organisations like KVIC can be involved in this activity. The building centres will also give a feedback to R & D organisations and promote the production of new materials. They will upgrade old skills and impart new skills to artisans. Building centres can conduct seminars/ workshops for architects and structural engineers in collaboration with NBO and regional research centres. The Board will have to play a crucial role in helping the entrepreneurs to set up production units with the assistance of financial institutions and making sure that government construction agencies place orders to create markets for new products. Even some of the Government's construction schemes can be linked up with the building centres to give a multiplier/ demonstration effect to the extension of new building skills/new materials.

To promote the production of building materials in a coordinated manner, the concept of building materials estates with common testing and equipment facilities needs to be promoted.

5   Recommendations

● It is obvious from the foregoing discussion that the adoption of capital and energy-intensive technologies (as is being practised today) cannot be expected to relieve the shortage of building materials without leading to environmental degradation. Any consideration of the choice of materials/technology must also take into account issues such as use of local resources and skills, employment generation, energy conservation and protection of environment.

● Any thrust in the reduction of building materials cost must come through:

- ○ Alternative wall-construction technologies
- ○ New developments in light-weight cheap roofings
- ○ Production of local cements
- ○ Development of components/materials which will replace or substantially cut down the consumption of timber, cement, steel etc.
- ○ Use of plastic materials in construction

● In view of the criteria laid out in the first paragraph, as part of the new thrust in building research, there has to be a strong emphasis on the development and adoption of the following new/alternative building materials:

○ Compacted stabilised soil blocks for wall construction
○ Polymer based coatings for mud, bamboo and thatch
○ Local production of lime-pozzolana cements, and other binders for soil stabilisation and plastering
○ Stabilised soil and improved locally burnt tiles for roof covering
○ Prefabricated roofs of bricks, tiles, ferrocement, precast RCC components (intermediate level of prefabrication at site), concrete roof tiles
○ Development of fibre reinforced materials using synthetic and natural fibres in cement and polymer matrices
○ Use of plastic pipes, tanks and fittings.

● The following long-term strategies must be adopted to conserve energy and conventional building materials:

○ Discourage use of ordinary Portland cement by maintaining substantial price difference between OPC and PPC
○ Convert all wet process cement plants into dry process ones
○ Modernise lime kilns to reduce energy consumption
○ Launch plantation programmes for timber, bamboo and other species
○ Use secondary timber and local fast growing trees
○ Increase productivity of timber and improved processing
○ Make designs using load-bearing pillars and footings.

● The Government may grant necessary fiscal and other concessions to building material manufacturing units utilising industrial and agricultural wastes. Similar concessions also should be given for thermosetting resins/FRP products, used for the manufacture of building components.

The sale price of the materials like bamboo should be manipulated

Table 5

A List of Development Activities to be Undertaken by Existing R&D Organisations

Material	Development Activity	Organisation
● Biomass (Bamboo/ Timber)	1. Plantation Programmes	Integrate with Wastelands Mission
	- Rural Needs (Social Forestry)	- Community lands/ Farmers Co-operatives
	- Industrial Needs (Production Forestry)	- Private companies on wastelands (on profit sharing basis)
	- Conservation of Forests (Ecological Forestry)	- Government Forest Departments
	2. Tissue culture techniques for propagation	- NCL, TERI, BARC, NBRI, IISc., JNU, Government Forest Departments, Forest Development Corporations (AICRP)
	3. Treatment of Bamboo /Timber to avoid decay and increase the life/ performance	- Forest Research Institute and Colleges, Dehradun
		- Forestry Research Institute, (Jabalpur)
		- Research Wing of Forest Departments (Maharashtra & U.P) and CBRI
	4. Improved Processing of Timber	- Forest Research Institute and Colleges, Dehradun

Table 5 (*Contd.*)

Material	Development Activity	Organisation
	5. Thatch Treatment	- CBRI, RRLs, IICT
	6. Timber like products (Boards/Panels) from alternative materials in polymer matrix	- NCL, CBRI, RRL Trivandrum, IICT, IPIRI
		- Private organisations (for instance WIP, IPM, NPL)
● Bricks, Blocks, Tiles	1. Soil stabilisation and soil stabilised mud blocks	- ASTRA, HUDCO, Development Alternatives, RECs, IITs (AICRP)
	2. Water proofing of mud walls/blocks	- IICT, IIP/IPCL, NCL
	3. Evaluation of soft mud presses (Modified version of MERADO machine)	- HUDCO, ASTRA, Development Alternatives, CBRI
	4. Evaluation of sand line/calcium silicate bricks	- HUDCO, ASTRA, Development Alternatives, CBRI
	5. Development of resin bonded sand tiles	- NCL, CSMRS, CSRL
	6. Development of bricks from industrial,	- NCB, CBRI, CFRI, CGCRI, RECs,

Table 5 (*Contd.*)

Material	Development Activity	Organisation
	agricultural and mineral wastes	IITs, Selected local engineering colleges (AICRP)
	7. Development of bricks from inferior soils	- CBRI, RRLs, RECs, IITs, Selected local engineering colleges (AICRP)
	8. Thermally efficient brick kilns	- CBRI, NCB, Private industry (Design organisations)
	9. Laterite blocks / tiles	- RRL- Trivandrum, CBRI
	10. Thermally efficient process for the manufacture of ceramic tiles	- CGCRI, CBRI, NCB, Private industry (Design organisations)
● Stone	1 Improved extraction/ processing techniques	- NCB, CBRI, REC - Jaipur, REC -Warangal, REC-Srinivasnagar
	2. Roofing / Walling systems	- NCB, CBRI, SERC, REC-Jaipur, REC-Warangal, REC-Srinivasnagar
● Prefabricated cement/ concrete products	1. Light weight expanded clay aggregates/cellular concrete (from waste materials)	- NCB, CBRI, IITs, CMERI

Table 5 (*Contd.*)

Material	Development Activity	Organisation
● Binders	1. Manufacture of different grades	- NCB, CBRI, ACC, CCI, DISIR, OCL
	2. Manufacture of blended cements	- NCB, ACC, CCI, DISIR, OCL
	3. Guidelines for use of blended cements and different grades	- BIS, NCB, CBRI
	4. Conversion of wet process cement plants into dry process	- DGTD, NCB
	5. Lime - Pozzolana	- KVIC, CBRI
	6. Development of binder/mansonry mortars from waste materials	- NCB, ACC, DISIR, CSRL, KVIC, CBRI
	7. Development of hydrated lime from other calcareous materials	- NCB, ACC, DISIR, CSRL, KVIC, CBRI
	8. Thermally efficient lime kilns	- CBRI, NCBM, Private Industries (Design organisation
	9. Gypsum based products	- CGCRI, SPIC, CBRI, RRL-Jammu

Table 5 (*Contd.*)

Material	Development Activity	Organisation
● Roofing Materials	1. Prefabricated framework for biomass infills	- ASTRA and other voluntary agencies to be identified by CAPART
	2. Nail jointed timber trusses or joinery made of ferrocement bandages	- FRIC, SERC
	3. Roofing sheets/ timber like products from alternative materials in cement matrix	- NCB, CBRI, RRL-Bhopal, CGCRI, SERC
		- Private organisations (for instance, Everest Building Products, Hyderabad Industries)
	4. Ferrocement products	- SERC, REC's, IIT's, CBRI, (AICRP)
	5. Concrete roof tiles	- NCB, CBRI
	6. Reinforced brick/ stone panels	- SERC, CBRI
	7. Upgradation of potter's and Mangalore tiles	- CGCRI, CBRI, KVIC

Table 5 (*Contd.*)

Material	Development Activity	Organisation
● Plastics for Building & Construction	1. Development of toilet pots, glazing, panels, tiles, roofing sheets etc.	- IPCL, CIPET, BIS, Plastic processors processors
	2. Studies on flammability of plastic materials and development of fire retardant grades.	- NCL, IPCL, IICT, CBRI

through fiscal adjustments keeping in view their use as primary building materials.

The major challenge in the rural housing sector is to increase the performance of traditional construction materials, particularly of those based on local renewable or abundant resources. Inputs of modern materials science can upgrade them, but this also increases the cost. It is suggested that subsidies may be granted under the housing programmes of the Government.

● Research and development in building materials has to be decentralised. Housing extension agencies must be set up on the lines of agriculture extension agencies. R & D must be integrated with demonstration units for production as well as application. Institutions geared to applied research in building technologies and their dissemination must be rejuvenated and networked suitably. Table 5 gives the details of development activities which can be taken up by different organisations based on their existing expertise. An institutional network, coordinated by the Building Materials Development Board, for the development of building materials and technologies is proposed. The *Building Centres* concept, being propagated by HUDCO and the Ministry of Urban Development, must be supported since it will fill the gap between research organisations and the construction sector and will act as extension agencies.

6   Acknowledgement

The inputs for this report received from Dr R K Bhandari, Director, CBRI; Prof S K Chopra, Former Additional Director General of Cement Research Institute (now NCB); Prof K S Jagadish, Convener, ASTRA Centre, IISc., Dr Ashok Khosla of Development Alternatives, Mr S K Sharma, Chairman and Managing Director, HUDCO, and other experts are greatly appreciated.

Appendix-1

List of abbreviations used in the text

ACC   Associated Cement Companies Limited
ASTRA   Centre for the Application of Science & Technology to Rural Areas
BARC   Bhabha Atomic Research Centre
BIS   Bureau of Indian Standards
CAPART   Council For Advancement of People's Action and Rural Technology
CBRI   Central Building Research Institute
CCI   Cement Corporation of India Limited
CFRI   Central Fuel Research Institute
CGCRI   Central Glass and Ceramic Research Institute
CIPET   Central Institute for Plastic Engineering and Tools
CMERI   Central Mechanical Engineering Research Institute
CSMRS   Central Soil and Materials Research Station
CSRL   Concrete and Soil Research Laboratory
DGTD   Directorate General of Technical Development
DISIR   Dalmia Institute of Scientific and Industrial Research
FRIC   Forest Research Institute and Colleges
FRI   Forestry Research Institute
GSI   Geological Survey of India
HUDCO   Housing and Urban Development Corporation
IICT   Indian Institute of Chemical Technology
IIP   Indian Institute of Petroleum
IISc   Indian Institute of Science
IIT   Indian Institute of Technology
IPCL   Indian Petrochemicals Corporation Limited
IPIRI   Indian Plywood Industries Research Institute
IPM   The Indian Plywood Manufacturing Company Limited
JNU   Jawaharlal Nehru University
KVIC   Khadi and Village Industries Commission
MERADO   Mechanical Engineering Research and Development Organisation
NABARD   National Bank for Agriculture and Rural Development
NBO   National Buildings Organisation
NBRI   National Botanical Research Institute

NCB	National Council for Cement and Building Materials
NCL	National Chemical Laboratory
NPL	Nuchem Plastics Limited
OCL	Orissa Cements Limited
PWD	Public Works Department
REC	Regional Engineering College
RRL	Regional Research Laboratory
SERC	Structural Engineering Research Centre
SPIC	Southern Petrochemical Industries Corporation Limited
TERI	Tata Energy Research Institute
WIP	The Western India Plywood Limited

16

Status on
National Technology Missions

The National Technology Missions are programmes
that, as a package, are meant to speed up
development. They break up the processes of change
and delivery into manageable tasks, and ensure that
ambitious deadlines are met as far as possible.

Contents

16
Status on
National Technology Missions

Preamble

The National Technology Missions are programmes that, as a package, are meant to speed up development. They break up the processes of change and delivery into manageable tasks that have ambitious deadlines. They focus on some key human needs. There are seven such missions relating to

- rural drinking water
- immunisation of pregnant women amd infants
- adult literacy
- self-sufficiency in edible oils
- improving the telecommunications network
- dairy development, and
- wasteland development.   ■

1   Introduction

Launched between 1985 and 1989, the Technology Missions aim to:

- significantly improve the availability and quality of drinking water in rural areas
- immunise infants against six grievous diseases and pregnant women against tetanus
- make a substantial proportion of the population functionally literate

793

- cut down import on edible oils
- extend, improve and innovate telecommunications
- improve milk production and rural employment
- bring under productive use the wastelands through a massive programme of afforestation and tree planting.

The missions emphasise:

- effective management
- self-reliance and self-help
- good quality and delivery of services
- reliable information and monitoring
- improved communications
- effective Centre-State coordination, and motivated
- community participation.

The background and critical issues of the Technology Missions and the significant achievements of the first five missions are described in this report. Reports on dairy development and wasteland development do not figure here since they are too recent to report.

1.1   Background and Critical Issues

These missions were selected based on inputs received from the science and technology community in the country. Based on this input, and the need to bring the benefits of science and technology to the masses, five missions, with a significant technological content, were selected during the Seventh Five Year Plan. It was well-recognised that these were probably not the five most important ones; but they were certainly critical for the overall national development.

The Missions are structured as part of the existing government ministries headed by individual Mission Directors, reporting to the Secretaries, who, in turn, report to the Ministers.

Since four out of the five missions are to be implemented at the state level, each state government has been requested to provide (a) a full-time Mission Director, for each individual mission, along with a senior officer, for mission coordination, and (b) appropriate Technology Mission Advisory Councils at the state level. The Councils report to the Chief Minister, and to the District Magistrate or equivalent at the district level.

The main purpose of these missions is to create a sense of urgency and provide missionary zeal, with the appropriate management focus, to deliver development which has been delayed for quite some time. The missions are also designed to identify linkages between various government agencies and departments and resolve bottlenecks for speedy implementation.

As pointed out earlier, there are at least five critical issues involved in running the technology missions: (a) management, (b) communication, (c) information, (d) Centre-State coordination and (e) people's participation. We shall briefly discuss each of these attributes.

1.1a   Management

Here, the need is to define objectives with a clear understanding of measurable milestones, and tangible quantifiable deliverables. It is also essential to define accountability, with autonomy and flexibility, to implement the different programmes.

1.1b   Communication

In a massive programme such as this one, affecting every corner of the country in one form or the other, effective communication to inform and implement various projects is essential. Communication on "what? why? how? when?" becomes the key to giving a sense of direction and purpose to many of these programmes. Communication at the national level involves the effective use of TV, radio, newspaper, newsletters, bulletins, seminars, workshops etc.

1.1c   Information

Information is essential for involvement and implementation. Information is also essential for effective feedback on performance. The information available in today's environment needs to be organised, formated, standardised and distributed openly. In fact, the information available today is generally obsolete, inaccurate and sometimes even fabricated. Setting up a management information system (MIS) for Technology Missions is a massive task involving (ultimately) computerisation at the district level for connectivity at the national level through NICNET. It could even begin with manual data, which may have some inaccuracies, to ultimately evolve into a viable MIS after a couple of years.

1.1d   Centre-State Coordination

As mentioned earlier, since four out of the five missions are implemented at the state level, effective communication with the state becomes a very critical parameter in implementation. For this, all of the national objectives outlined in the national document are being further divided into state level objectives and documents. It is hoped that state level objectives will be further divided into district level goals for effective micro level implementation and planning. Periodic interactions with the state is essential to motivate the state level machinery for these missionary tasks.

1.1e   People's Participation

Since these are *national* technology missions, the people's participation is critical for effective implementation. Without a massive people's participation, involving voluntary agencies, academicians, professionals, private and public sectors, universities, retired officers etc., these missions cannot be implemented in totality. For example, we organised a workshop in Anand to explore the role that dairy cooperatives can play in these missions. Since dairy cooperatives are already into 51000 villages, we believe that it is feasible to utilise their network for delivering immunisation and literacy programmes. This is really the crux of it all: involve as many groups of people as possible for motivation, education, awareness and feedback.

In order to operationalise these missions, a clear focus on technological self-reliance was desirable to expedite development and provide long-term maintainability. In each of these areas various technological capabilities are available in our laboratories and the task is to bring these techniques from the laboratories to the land, through effective prototyping, field trials and mass production. In order to operationalise the delivery system, it was essential to set up parallel channels to ensure effective development.

In the following pages we will briefly describe the achievements and progress of the five technology missions.

2   Drinking Water

Some of the achievements of this mission are outlined below:

● Out of the 1,61,722 'no source' problem villages, as on 1 April 1985, 1,43,298 problem villages have been covered till 31 August 1989. Of the remaining 18,424 villages, 14,229 will be covered during the remaining period of the Seventh Plan and 4195 will be carried over to the Eighth Plan.

● Ground water potential maps were completed for all the 55 mini mission districts.

● Sources were identified in 8584 villages, upto August 1989, by the Central Ground Water Board, out of the 13,187 problem villages indicated to them. The remaining are likely to be identified by 31 March 1990.

● 120 SPV pumping systems are to be set up by 31 March 1990 for drawal of water in remote villages without conventional energy. Six such systems have already been installed.

● 5000 iron removal plants are to be set up in 1989-90. 600 have already been installed.

● 128 desalination plants are to be set up under the Sub Mission on Desalination of Water. 14 such plants have already been set up.

● Out of the 130 defluoridation plants, to be set up under the Sub Mission on Control of Fluorosis, 5 have already been installed and the remaining will be completed by 31 March 1990.

● The number of guineaworm affected villages has been reduced to 3111. The Mission will achieve the target of eradication of guinea-worm by the end of the Seventh Plan. Against the target of conversion of 2892 step wells into sanitary wells, 700 wells have already been constructed.

● The progress of the coverage of problem villages with safe drinking water facilities is being monitored in terms of their names.

● The computerised rig monitoring system has been introduced for optimum utilisation of the capacity of the rigs used for drilling.

● 87 stationary district level laboratories and 17 mobile water testing laboratories are to be set up by 31 March 1990. 7 such laboratories have already started functioning.

● The earmarking of funds for water conservation, under RLEGP (now JRY), has been carried out.

● For the first time, quality standards have been formulated and prescribed for tubewells.

● The states and union territories have been advised to set up village, district and state level water programme review committees.

3   Immunisation

With a concerted effort, the coverage levels of various vaccines (DPT, BCG, Polio) increased from the 29-41 per cent level to 55-79 per cent throughout the country.

In order to protect the efficacy of the vaccine during transit, from the manufacturer to the consumer, appropriate cold-chain infrastructures were provided (104 walk-in-coolers, 3217 refrigerators, 1632 chest freezers and chest refrigerators, 34,000 cold boxes and an equal number of vaccine carriers).

Some of the other achievements of this technology mission include:

● Cold chain systems were indigenised. Indigenously developed ice-line refrigerators and cold-boxes carriers are being field tested in 40 districts.

● An indigenous measles vaccine was introduced in the immunisation programme. Plans for the oral polio vaccine and killed polio vaccine were finalised and manufacturing facilities are being set up.

● 30 medical colleges are being associated actively with the surveillance and epidemeiological analysis in 30 districts, along the lines of the North-Arcot model.

● Nine vaccine testing facilities have now been extended to other institutions; there was only one such facility at the start of the mission.

● Evaluation surveys were made a regular feature both at regional and national levels; 291 surveys were carried out, these surveys also considered the qualitative aspects of the service delivery.

● Reach of the services was also extended to urban areas. Special action plan have been drawn up for 12 metropolitan cities.

Professional societies like the Indian Academy of Paediatrics, the Indian Medical Association and many voluntary agencies are also involved in the programme. This has helped in mobilising community support to the programme.

All medical colleges in the country have been involved in the programme, from the planning to the evaluation stage, within their field practice areas, thereby inculcating a sense of community service in the students.

Out-break investigations were made an integral part of the disease surveillance system. Extensive disease incidence data is being obtained through sentinel information systems.

Programme planning and monitoring is being done through a management information system, making the implementation easy. Nonformal pressure groups are also being activated for generating demand for the services.

4 Literacy

4.1 Highlights of the Mission

The following are the highlights of the national literacy mission:

● A mass campaign launched by the Prime Minister on 5 May 1988. A similar campaign was launched by 24 States and Union Territories on the same date and after.

● The National Literacy Mission Authority (NLMA) with a Council and Executive Committee was constituted. The Executive Committee met 13 times and took several important decisions.

● State Literacy Mission Authorities (SLMA) were constituted in 18 states.

● The Saksharta Abhiyan programme was launched by the Gujarat Vidyapeeth on 1 May 1988, with 416 voluntary agencies and 1.5 lakh volunteers, to make 35 lakh adult illiterates literate in five phases by 1991. Employers, trade unions, universities, colleges, school teachers, students, non-student youth, banks, cooperatives, Rotary and Lion Clubs, and Jaycees were involved in the Abhiyan.

● A mass campaign for complete eradication of illiteracy was launched in 20 taluks of Karnataka in November 1988. 40,000 volunteers are actively engaged in imparting literacy to make four lakh adult illiterates literate by 1990. An additional 40 taluks in five districts are being taken up, in the second phase of the campaign, to make 13.5 lakh illiterates literate.

● 7,000 students of the progressive Delhi schools are working for the NLM since September 1989. The number is likely to increase to 10,000 in 1989-90. An evaluation of the performance also is being conducted.

● 1,56,000 school students of Rajasthan are also involved in NLM from the summer of 1989. The required number of literacy kits have been distributed.

● The Governments of Orissa and West Bengal have taken a decision to compulsorily involve the entire student community, from Class IX onwards, in the NLM. Necessary directions have been issued to the Boards of Secondary Education.

● A mass mobilisation campaign was launched, under the auspices of the NSS Wing of Mahatma Gandhi University, Kottayam, to make 2000 illiterates in Kottayam city fully literate within 100 days. Kottayam city was declared fully literate on 25 June 1989.

● A mass mobilisation campaign was also launched for the complete eradication of illiteracy in Ernakulam, under the auspices of Kerala Shastra Sahitaya Parishad and under the leadership of Collector, to make 2.5 lakh illiterates in Ernakulam district fully literate. School teachers, students, unemployed and employed youth, retired employees, retired school teachers, police officers, medical officers, bank officers, trade unions, priests and nuns, religious organisations (church, mosque and temples), the Nair Service Society etc. were fully involved in this effort. 70 % of the 21,000 instructors are girls.

● Plans for the complete eradication of illiteracy in Kerala and Pondicherry have been approved and communicated.

● Plans for the complete eradication of illiteracy in 20 Blocks of West Bengal were approved and communicated. The plan was launched on 8 September, 1989.

● The Government of West Bengal has resolved to make West Bengal fully literate by 1995, the 175th anniversary of Pandit

Ishwarchand Vidyasagar. All the constituents of the Left Front Government were involved in the campaign to make 9 million adult illiterates in West Bengal literate.

● The plan for the complete eradication of illiteracy in the Coimbatore district was launched in May 1988. District level coordination committees were formed; 30 institutions and voluntary agencies and 15,000 school students were involved in the project and 10 villages have so far been made fully literate.

● 18 villages in Rajasthan have been made fully literate under the whole village literacy concept between 1987-89.

● 30,000 volunteers were mobilised in Bombay University to convert at least one lakh adult illiterates by 1990. An exciting experiment, of a 'human chain' over an area of 11 km in Bombay, is being launched by Dr Madhav Chavan of Bombay Institute of Chemical Technology.

4.2 Participation of Different Agencies

Several institutions, services and agencies are involved in the national literacy mission. The involvement of some of these agencies is discussed below:

Army, Navy and Air Force and Their Welfare Organisations

The Army, Navy and Air Force Wives' Welfare Associations at Cochin, Visakhapatnam, New Delhi and Lucknow are actively involved in literacy work.

Naval Headquarters issued detailed instructions to all the three naval commands urging them to take up literacy work for the benefit of the familiies of service and civilian personnel and domestic servants.

Ex-Servicemen

50 blocks in the six states of Bihar, Haryana, Himachal Pradesh,

Madhya Pradesh, Rajasthan and Tamil Nadu have been finalised for involvement of ex-servicemen. Grants have been released in favour of 44 projects (each project of 300 centres) in six States.

Railways

The Railway Board implemented 425 centres, on 15 August 1988, and enrolled 11,362 persons in nine zones. In 1989-90, the number of centres has gone up to 600.

Prison Management

The prison management and staff is actively working to impart literacy to the life convicts in the jails of West Bengal, Madhya Pradesh, Andhra Pradesh and Rajasthan. The Union Home Secretary has written to Chief Secretaries of States/Union Territories about this invovement.

Banks

The Director General(NLM) wrote to the Chairmen and Managing Directors of the 20 nationalised banks. State Bank of India, Allahabad Bank, Syndicate Bank, Canara Bank and Andhra Bank have already agreed to allow their employees to participate in the missions, and identified such employees.

Cooperatives

The National Cooperative Union of India issued directions to all State Cooperative Unions of India to allow the literate members of the cooperatives to impart literacy.

Voluntary Agencies

750 projects, involving 43,000 adult education centres, have been sanctioned to 551 voluntary agencies during 1987-88 and 1988-89. A joint evaluation of the performance of 346 agencies was conducted in 87-88. The joint evaluation repeated for Andhra Pradesh during 1989-90.

4.3   New and Integrated Technique to Impart Literacy

The acquisition and retention of different learners being not uniform, three NLM primers, representing three different levels of skills in literacy and numeracy, are being designed to incorporate work book, exercise book, tools of evaluation of the learning outcome, certificate to the learner etc. The overall duration of the learning is being sought to be reduced to 200 hours or about 6 months under this integrated technique known called *improved pace and content of learning,* or simply IPCL. An expert group has been constituted at the national level for the scrutiny of the primers.

The involvement of spiritual leaders like Baba Amte, Acharya Tulsi, Panduranga Athavale and Swami Ranganatha Nanda is also being sought for the mission.

4.4   Integration of Literacy with Development

After a review of the pace and progress of the Mission by the Cabinet Secretary on 15 December 1988, detailed instructions have been issued by Chief Secretaries of most of the State Governments to their development functionaries to view literacy as their own programme. This involvement appears to be slowly and gradually gathering momentum.

5   Oilseeds

The Oilseed Mission, launched in May 1987, with the objective of accelerating self-reliance in edible oils, has made great strides. Imports have been cut down by 80%; in fact, India now exports oilseeds products.

40 new varieties of different oilseed crops have been developed; the yield potential of these is twice the national average, under farmers conditions, and four times under research farm conditions.

The production of breeder seeds made a quantum jump from 2762 quintals to 8094 quintals in 1988-89, an increase of 193% during the three years of the Mission.

CSIR has developed a batch type process for rice bran stabilisation and sunflower decortication; its demonstration is on, and the technology is being made available to manufacturers and users. An improved expeller has also been developed, and is currently being commercialised.

A thrust programme was taken up in 246 districts to demonstrate profitable technology and provide inputs like seeds, bio-fertilisers, gypsum, plant protection etc. to farmers. The investment for this programme has been raised from Rs.30 crores, before the start of mission, to Rs.67 crores in 1988-89.

An incentive price for groundnut and rapeseed has been ensured through NDDB's procurement policy. The consumer can now also get edible oil at a reasonable price during the lean season.

The cooperative system, and its storage and processing facilities have been expanded. A modification of huller mills has helped to make available a large quantity of bran for oil extraction.

Apart from a good monsoon, other factors, like the confidence of oilseeds farmers to use technology and realise profits with incentive prices, contributed to this breakthrough in the oilseeds sector.

6   Telecom

The Telecom Mission, launched in April 1986, has made significant achievements in the following activities:

● The quality of service has been improved. Local/STD calls success rates increased, and the number of faults were reduced considerably through better maintenance, by replacing worn-out equipment and by streamlining the procedures for fault analysis and other customer related activities.

- The availability of urban PCOs increased: local 15,6000, trunk 5700 and STD 900. About 150 telecom centres provided facilities of PCOs, local/trunk/STD, telex and fax under a single roof.

- The delivery of telegrams improved. 82% telegrams are now being delivered within 12 daylight hours, as against 29% in April 1986. New technology for electronic key boards, key board concentrators, store and forward message switching system, is being inducted.

- Telex connections are being made available on demand in many of the stations; a net capacity of 8500 lines has been added. The waiting list was brought down to 1800 (as against 2800 in April 1986). Seventy notional telex exchanges were also opened to meet requirements in remote areas.

- Of the 50,421 inhabitated hexagons, 31,000 were provided with telecom facilities. More than 100 electronic exchanges were inducted in the rural areas.

- A national digital network is being built. Of the 447 district headquarters, 324 were provided with STD facilities. The four metros are being connected on digital media. Six digital trunk automatic exchanges were also commissioned to improve inter-city telecom.

Annexures

Annexure-1

Composition and Terms of Reference of Science Advisory Council to the Prime Minister*

Professor C N R Rao Director Indian Institute of Science Bangalore-560 012	*Chairman*
Professor Madhav Gadgil Centre for Ecological Sciences Indian Institute of Science Bangalore-560 012	*Member*
Professor R Narasimha Director National Aeronautical Laboratory P B No 1779, Kodihalli Bangalore-560 017	*Member*
Dr R A Mashelkar Director National Chemical Laboratory Pune-411 008	*Member*
Dr Sekhar Raha Chief Executive ICI India Limited 3rd Floor, Ashoka Annexe Chanakya Puri New Delhi-110 021	*Member*
Professor V L Chopra Professor of Eminence & Head Biotechnology Centre Indian Agricultural Research Institute New Delhi-110 012	*Member*

*Constituted by Cabinet Secretariat Notification, F No A-11019/1/86/Ad 1 dated February 4, 1986

Professor J V Narlikar	*Member*
Director and Professor	
Inter University Centre for	
Astronomy and Astrophysics	
Poona University Campus	
Pune-411 007	

Dr P Rama Rao   *Member*
Director
Defence Metallurgical Research Laboratory
Kanchanbagh P O
Hyderabad-500 258

Dr A S Ganguly   *Member*
Chairman
Hindustan Lever Limited
Hindustan Lever House
Backbay Reclamation
Bombay-400 020

S G Pitroda   *Member*
Adviser to PM on
Technology Missions
9th Floor, Akbar Bhavan
Chankya Puri
New Delhi-110 021

Professor P N Tandon   *Member*
Professor of Neurosurgery
Neurosurgery Centre
All India Institute of Medical Sciences
Ansari Nagar
New Delhi-110 029

Dr P J Lavakare   *Secretary*
Adviser   *to the Council*
Department of Science & Technology
Technology Bhavan
New Mehrauli Road
New Delhi-110 016

Terms of Reference

The Council will advise the Prime Minister on:

- Major issues facing Science and Technology today
- The health of Science and Technology in the country and the direction in which it should move and
- A perspective plan for A D 2001.

The Council will also look at specific problems with different scientific departments, policies, priorities for research and technology missions, etc.

Annexure-2

Composition of Sub-Committees Set Up to Prepare the Technical Reports

Advanced Materials:
National Priorities

Dr P Rama Rao
Director
Defence Metallurgical Research
Laboratory
Kanchabagh P O
Hyderabad-500 258

Professor S Ranganathan
Professor & Chairman
Department of Metallurgy
Indian Institute of Science
Bangalore-560 012

Professor K J Rao
Department of Metallurgy
Indian Institute of Science
Bangalore-560 012

Photonics

Professor J V Narlikar
Director
Inter-Univerisity Centre
for Astronomy and Astrophysics
Poona Univeristy Campus
Pune-411 007

Professor S V Pappu
Department of Electrical
Communication Engineering
Indian Institute of Science
Bangalore-560 012

Dr A K Sreedhar
Director
Solid State Physical Laboratory
Lucknow Road
Delhi-110 007

Professor C M Srivastava
Head
Advanced Centre for Research
in Electronics
Indian Institute of Technology
Bombay-400 076

Professor B V Sreekantan
Director
Tata Institute of Fundamental
Research
Homi Bhabha Road
Bombay-400 005

Dr D D Bhawalkar
Director
Centre for Advanced Technology
Department of Atomic Energy
P O Rajendra Nagar
Indore-452 012

Professor R Vijayaraghavan
Head
Solid State Physics and
Materials Science Group
Tata Institute of Fundamental
Research
Homi Bhabha Road
Bombay-400 005

Priorities of Genetic Research in India

Dr A S Ganguly
Chairman
Hindustan Lever Limited
Hindustan Lever House
Backbay Reclamation
Bombay-400020

Dr K K G Menon
Vice-President (Research)
Hindustan Lever Research Centre
Andheri
Bombay-400 099

Dr Khorshed Pavri
Director
National Institute of Virology
20-A Ambedkar Road
Pune-411 007

Dr N K Notani
Director & Head
Biology Group
Bhabha Atomic Research Centre
Trombay
Bombay-400 085

Professor G Padmanabhan
Department of Biochemistry
Indian Institute of Science
Bangalore-560 012

Professor O Siddiqi
Professor of Molecular Biology
Tata Institute of Fundamental
Research
Homi Bhabha Road
Bombay-400 005

Professor S Modak
Department of Zoology
University of Pune
Pune-411 007

Lasers

Dr D D Bhawalkar
Director
Centre for Advanced Technology
Department of Atomic Energy
P O Rajendra Nagar
Indore-452 012

Parallel Computing

Professor V Rajaraman
Head, Computer Centre
Indian Institute of Science
Bangalore-560 012

Professor S Ramani
Computer Centre
Tata Institute of Fundamental
Research
Homi Bhabha Road
Bombay-400 005

Professor P C P Bhat
Head
Computer Centre
Indian Institute of Technology
Hauz Khas
New Delhi-110 016

Dr P N Shankar
Scientist
Fluid Mechanics Division
National Aeronautical Laboratory
Bangalore 560 017

Dr N Seshagiri
Additional Secretary
Department of Electronics
Lok Nayak Bhavan
Khan Market
New Delhi-110 003

Instrumentation

Professor P N Tandon
Professor of Neurosurgery
Department of Neurosurgery
All India Institute of Medical
Sciences
Ansari Nagar
New Delhi-110 029

Dr G Venkataraman
Indira Gandhi Centre
for Atomic Research
Kalpakkam
Madras-603 102

Dr R Hardaynath
Director
IRDE
Dehra Dun

Dr S R Gowariker
Director
Central Scientific Instruments
Organisation
Chandigarh-160 020

A K Verma
Engineers India Limited
Bhikaji Kama Place
R K Puram
New Delhi-110 022

Professor K V Sane
Department of Chemistry
University of Delhi
Delhi-110 007

Dr Ashok Khosla
22 Palam Marg, Vasant Vihar
New Delhi

Dr H C Verma
AIMJL, Naimex House
A-8, Mohan Cooperative
Industrial Estate, Mathura Road
New Delhi-110 044

Professor S K Guha
Biomedical Engineering
Research Centre
Indian Institute of Technology
New Delhi-110 016

C S Srinivasan
CEL 4, Industrial Area
Sahibabad-201 010

Dr R P Singh
Director
Department of Science and
Technology
New Delhi-110 016

**Robotics and Manufacturing
Automation**

N V Ramaswamy
Division of Remote Handling
and Robotics
Bhabha Atomic Research Centre
Bombay-400 085

Professor N Viswanadham
Computer Science & Automation
Indian Institute of Science
Bangalore-560 012

Dr V Venkateswarlu
Manager (Production)
C-DOT, 71 Sona Towers
Miller's Road
Bangalore-560 052

Professor A Ghosh
Department of Mechanical
Engineering
Indian Institute of Technology
Kanpur-208 016

Dr N R Mantena
General Manager
CNC Division, HMT Ltd
Bangalore-560 052

**Management of Renewable
Resources**

Professor Madhav Gadgil
Centre for Ecological Sciences
Indian Institute of Science
Bangalore-560 012

Dr R S Paroda
Deputy Director General
Indian Council of Agricultural
Research
New Delhi-110 001

Professor H Y Mohan Ram
Department of Botany
University of Delhi
Delhi-110 007

Dr V M Meher-Homji
Dean
Salim Ali School of Ecology
Pondicherry University
French Institute, P B No 33
Pondicherry-605 001

Dr Hari Narain
National Geophysical
Research Institute
Hyderabad-500 007

Dr S K Sinha
Professor of Eminence
Plant Physiology &
Water Technology Centre
Indian Agricultural
Research Institute
New Delhi-110 012

Professor J S Singh
Department of Botany
Banaras Hindu University
Varanasi-221 005

Health Care

Professor P N Tandon
Professor of Neurosurgery
Neurosurgery Centre
All India Institute of Medical
Sciences
New Delhi-110 029

Professor L M Nath
Professor & Head
Centre for Community Medicine
All India Institute of Medical
Sciences
New Delhi - 110 029

Professor N H Antia
Director
The Foundation for
Research in Community Health
84-A, R G Thadani Marg
Worli
Bombay-400 018

Professor Sneh Bhargava
Director
All India Institute of Medical
Sciences
New Delhi-110 029

Professor Ashish Bose
Professor & Head
Population Research Centre
Institute of Economic Growth
University of Delhi
Delhi-110 007

Professor Nirmala Murthy
Public Systems Group
Indian Institute Management
Vastrapur
Ahmedabad-380 015

Dr T P Sharma
Director
Public Health & Family Welfare
Directorate of Health Services
Bhopal

Professor K Srinivasan
Director
International Institute for
Population Sciences
Govindi Station Road
Deonar
Bombay-400 088

Fertilizer Use

Professor V L Chopra
Professor of Eminence & Head
Biotechnology Centre
Indian Agricultural Research
Institute
New Delhi-110 012

Dr K Kanungo
Member
Agricultural Scientists
Recruitment Board
Krishi Anusandhan Bhawan
New Delhi-110 012

Dr Rajendra Prasad
Head, Agronomy Division
Indian Agricultural Research
Institute
New Delhi-110 012

Dr N N Goswami
Dean & Joint Director (Education)
Indian Agricultural Research
Institute
New Delhi-110 012

Dr Prem Narain
Director
Indian Agricultural Statistical
Research Institute
Library Avenue
New Delhi-110 012

Dr G R Saini
Advisor (Economic & Statistical)
Department of Agriculture and
Cooperation, Krishi Bhawan
New Delhi-110 001

Dr H L S Tandon
Director
Fertilizer Development and
Consultation Organisation
C-110 Greater Kailash -I
New Delhi-110 048

Dr B C Biswas
Director
Fertilizer Association of Indian
New Mehrauli Road
New Delhi-110 016

Dr R P Singh
Director
Central Research Institute for
Dryland Agriculture
Hyderabad

Dr Virenda Kumar
Chief (Agriculture Services)
Indian Farmers Fertilisers
Cooperative Limited
53-54 Nehru Place
New Delhi-110 019

Dr S K Mukherjee
332 Jodh Park
Calcutta-700068

Dr K R Kulkarni
Project Coordinator
All India Coordinated
Agronomic Research Project
Indian Council of Agricultural
Research
University of Agricultural
Sciences
Bangalore-560 065

Future Food Needs

Professor V L Chopra
Professor of Eminence & Head
Biotechnology Centre
Indian Agricultural Research
Institute
New Delhi-110 012

Dr A S Ganguly
Chairman
Hindustan Lever Limited
Hindustan Lever House
Backbay Reclamation
Bombay-400 020

Dr M S Swaminathan
President
International Union for
Conservation of Nature and
Natural Resources
11 Rathna Nagar
Teynampet
Madras-600 018

Dr P N Bhat
Director
Indian Veterinary Research
Institute
Izat Nagar

Dr S N Dwivedi
Additional Secretary
Department of Ocean
Development
Central Goverment Office
Complex
Lodi Road
New Delhi-110 003

Dr S K Sinha
Professor of Eminence
Plant Physiology and
Water Technology Centre
Indian Agricultural Research
Institute
New Delhi-110 012

Dr I P Ibrol
Deputy Director General
Indian Council for Agricultural
Research, Krishi Bhavan
New Delhi-110 0C1

Ms Rami Chhabra
Advisor (Media & Communication)
Ministry of Health & Family
Welfare, Nirman Bhavan
New Delhi

**Chemical Industry in the year
2000 AD: National Priorities**

Dr R A Mashelkar
Director
National Chemical Laboratory
Pune-411 008

Professor M M Sharma
Department of Chemical
Technology, University of Bombay
Bombay-400 019

Water Transport in India

Professor J V Narlikar
Director
Inter-Univerisity Centre
for Astronomy and Astrophysics
Poona Univeristy Campus
Pune-411 007

M A R Ansari
Member (Technical)
Inland Waterways
Authority of India
B/109-112, Commercial
Complex, Sector-18
Noida

Xavier Arakal
Chairman
Inland Waterways Authority of
India
Transport Bhavan
1 Parliament Street
New Delhi-110 001

E J D'Sa
General Manager
The Shipping Corporation
of India Limited, Shipping House
245, Madame Cama Road
Bombay-400 021

M H Patwardhan
Essel's Amusement Parks (India)
Continental Building
135 Annie Besant Road
Bombay-400 018

Capt Augustine Rebello
Captain of Ports
Govt of Goa
Panaji

D Sanyal
Executive Director (in-charge)
National Transportation Planning
and Research Centre
Eekas Sadan, 78-A Zamrudpur
Greater Kailash Part I
New Delhi-110 048

R C Sharma
Director
Directorate of Transport Research
1DA Building
Jamnagar House
New Delhi-110 011

Dr Z S Tarapore
Director
Central Water Power Research
Station
Khadakwasla
Pune-411 024

V A Valiaparampil
Joint Adviser (Transport)
Planning Commission
Yojana Bhawan
New Delhi-110 001

Professor D J Victor
Professor of Civil Engineering
Indian Institute of Technology
Madras-600 036

Dr A B Wagh
Scientist
National Institute
of Oceanography
Dona Paula-403 004

Minerals Development

Dr P Rama Rao
Director
Defence Metallurgical Research
Laboratory
Kanchanbagh P O
Hyderabad-500 258

Mahadevan
Director
Atomic Minerals Division
Hyderabad

Professor V K Gaur
Secretary, DOD, CGO Complex
Lodi Road
New Delhi

Mr Mani
Adviser, Minerals
Minerals Development Board
Government of India
New Delhi

Building Materials

Dr R A Mashelkar
Director
National Chemical Laboratory
Pune-411 008

Professor Madhav Gadgil
Professor
Centre for Ecological Sciences
Indian Institute of Science
Bangalore-560 012

Dr R K Bhandari
Director
Central Building
Research Institute
Roorkee-247 667

Dr Ashok Khosla
Development Alternatives
22 Palam Marg, Vasant Vihar
New Delhi.